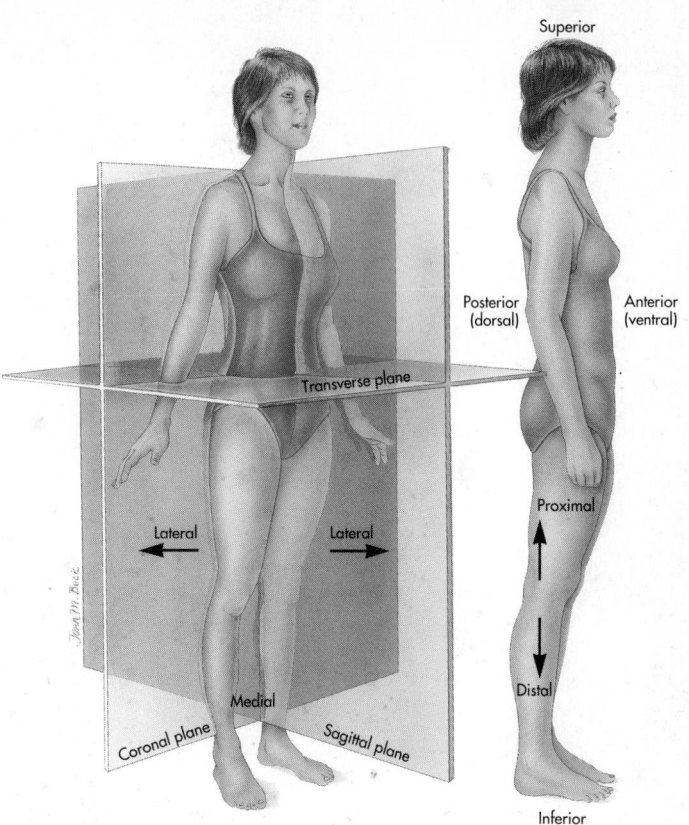

To make the reading of anatomical figures a little easier, an *anatomical compass* is used throughout this book. On many figures, you will notice a small compass rosette similar to those on geographical maps. Rather than being labeled N, S, E, and W, the anatomical rosette is labeled with abbreviated anatomical directions.

A = Anteri~~or~~
I = Inferior   ~~P~~ = ~~Posterior~~
L (opposite R) = Left   R = Right
L (opposite M) = Lateral   S = Superior

Pocket Reference to accompany

ANATOMY & PHYSIOLOGY

Third Edition

Gary A. Thibodeau
Chancellor and Professor of Biology
University of Wisconsin—River Falls
River Falls, Wisconsin

Kevin T. Patton
Professor, Department of Life Sciences
St. Charles County Community College
St. Peters, Missouri

 Mosby

St. Louis   Baltimore   Boston
Carlsbad   Chicago   Naples   New York   Philadelphia   Portland
London   Madrid   Mexico City   Singapore   Sydney   Tokyo   Toronto   Wiesbaden

Mosby

Dedicated to Publishing Excellence

A Times Mirror Company

Editor: Ron Worthington
Developmental Editor: Jean Fornango
Project Manager: Carol Sullivan Weis
Production Editor: Florence Achenbach
Design: Sheila Barrett

Printed in the United States.

Mosby–Year Book, Inc.
11830 Westline Industrial Drive, St. Louis, Missouri 63146

International Standard Book Number 0-8151-8829-3

Contents

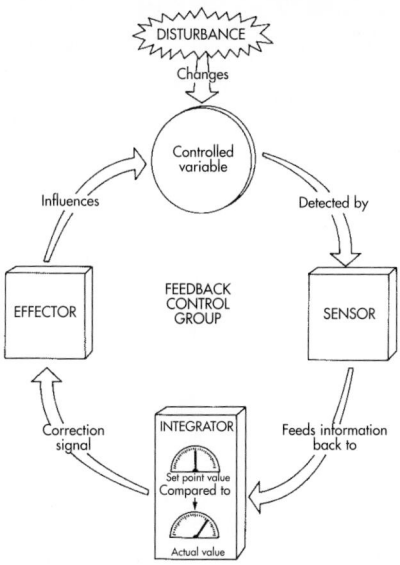

Basic components of homeostatic control mechanisms. A physiological variable such as blood pressure or body temperature is frequently disturbed by any various factors. To maintain relative stability, or homeostasis, of the variable, the body resorts to feedback control mechanisms that often follow the scheme illustrated here. One or more sensors, cells sensitive to changes in the controlled variable, detect changes in the variable and feed the information back to control centers called integrators. Integrators, perhaps in the central nervous system, collect actual information about the variable (from the sensors) and compare it to "built in" or preprogrammed information called the **set point** value. The integrator then sends an **error signal** (by way of hormones and/or nerve impulses) to an organ or group of organs called an **effector.** In response to the error signal, the effector produces an effect on the variable—generally influencing the variable to move back toward the set point value. Once the set point is reached, feedback from the sensors causes the integrator to stop the error signal.

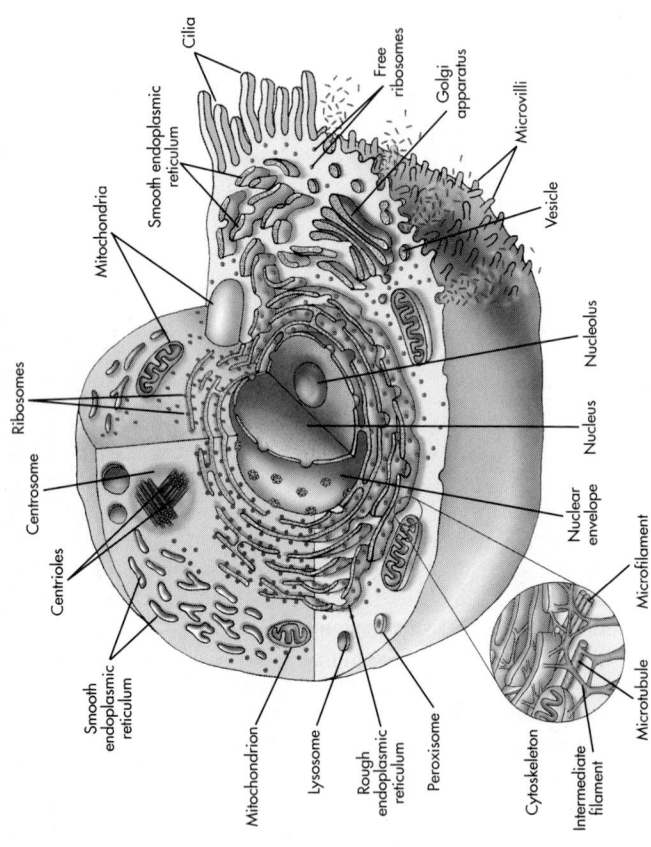

Cilia
Free ribosomes
Golgi apparatus
Microvilli
Smooth endoplasmic reticulum
Mitochondria
Vesicle
Ribosomes
Nucleolus
Centrosome
Nucleus
Centrioles
Nuclear envelope
Smooth endoplasmic reticulum
Mitochondrion
Lysosome
Rough endoplasmic reticulum
Peroxisome
Microfilament
Cytoskeleton
Microtubule
Intermediate filament

Typical or composite cell.

SOME MAJOR CELL STRUCTURES AND THEIR FUNCTIONS

Cell Structure	Functions
MEMBRANOUS	
Plasma membrane	Serves as the boundary of the cell, maintaining its integrity; protein molecules embedded in plasma membrane perform various functions; for example, they serve as markers that identify cells of each individual, as receptor molecules for certain hormones and other molecules, and as transport mechanisms
Endoplasmic reticulum (ER)	Ribosomes attached to rough ER synthesize proteins that leave cells via the Golgi complex; smooth ER synthesizes lipids incorporated in cell membranes, steroid hormones, and certain carbohydrates used to form glycoproteins
Golgi apparatus	Synthesizes carbohydrate, combines it with protein, and packages the product as globules of glycoprotein
Lysosomes	A cell's "digestive system"
Peroxisomes	Contain enzymes that detoxify harmful substances
Mitochondria	Catabolism; ATP synthesis; a cell's "power plants"
Nucleus	Houses the genetic code, which in turn dictates protein synthesis, thereby playing essential role in other cell activities, namely, cell transport, metabolism, and growth
NONMEMBRANOUS	
Ribosomes	Site of protein synthesis; a cell's "protein factories"
Cytoskeleton	Acts as a framework to support the cell and its organelles; functions in cell movement; forms cell extensions (microvilli, cilia, flagella)
Cilia and Flagella	Hairlike cell extensions that serve to move substances over the cell surface (cilia) or propel sperm cells (flagella)
Nucleolus	Plays an essential role in the formation of ribosomes

SOME IMPORTANT TRANSPORT PROCESSES

Process	Type	Description		Examples
Simple Diffusion	Passive	Movement of particles through the phospholipid bilayer or through channels from an area of high concentration to an area of low concentration—that is, down the concentration gradient		Movement of carbon dioxide out of all cells; movement of sodium ions into nerve cells as they conduct an impulse
Dialysis	Passive	Diffusion of small solute particles, but not larger solute particles, through a selectively permeable membrane; results in separation of large and small solutes		During procedure called *peritoneal dialysis,* small solutes diffuse from blood vessels but blood proteins do not (thus removing only small solutes from the blood)
Osmosis	Passive	Diffusion of water through a selectively permeable membrane in the presence of at least one impermeant solute		Diffusion of water molecules into and out of cells to correct imbalances in water concentration
Facilitated Diffusion	Passive	Diffusion of particles through a membrane by means of carrier molecules; also called *carrier-mediated passive transport*		Movement of glucose molecules into most cells

Active Transport	Active	Movement of solute particles from an area of low concentration to an area of high concentration (up the concentration gradient) by means of a carrier molecule		In muscle cells, pumping of nearly all calcium ions to special compartments—or out of the cell
Phagocytosis	Active	Movement of cells or other large particles into a cell by trapping it in a section of plasma membrane that pinches off to form an intracellular vesicle; type of *endocytosis*		Trapping of bacterial cells by phagocytic white blood cells
Pinocytosis	Active	Movement of fluid and dissolved molecules into a cell by trapping them in a section of plasma membrane that pinches off to form an intracellular vesicle; type of *endocytosis*		Trapping of large protein molecules by some body cells
Exocytosis	Active	Movement of proteins or other cell products out of the cell by fusing a secretory vesicle with the plasma membrane		Secretion of the hormone, prolactin, by pituitary cells

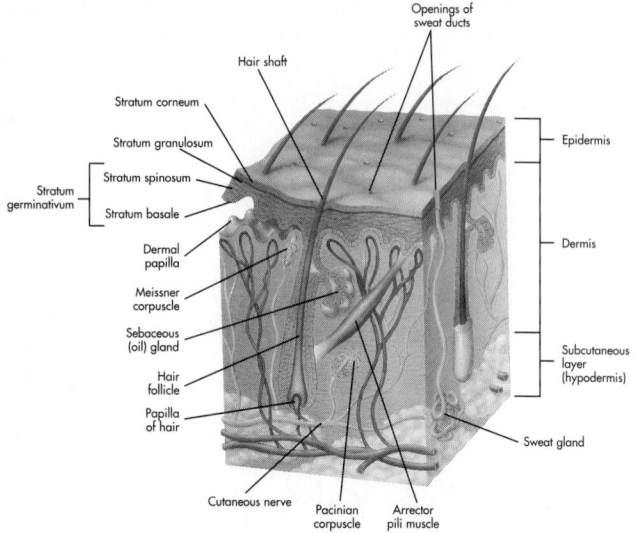

Microscopic diagram of the skin.

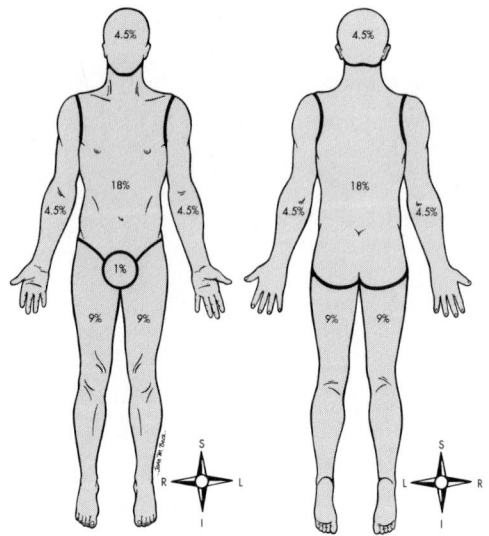

"Rule of nines: is one method used to estimate amount of skin surface burned in an adult.

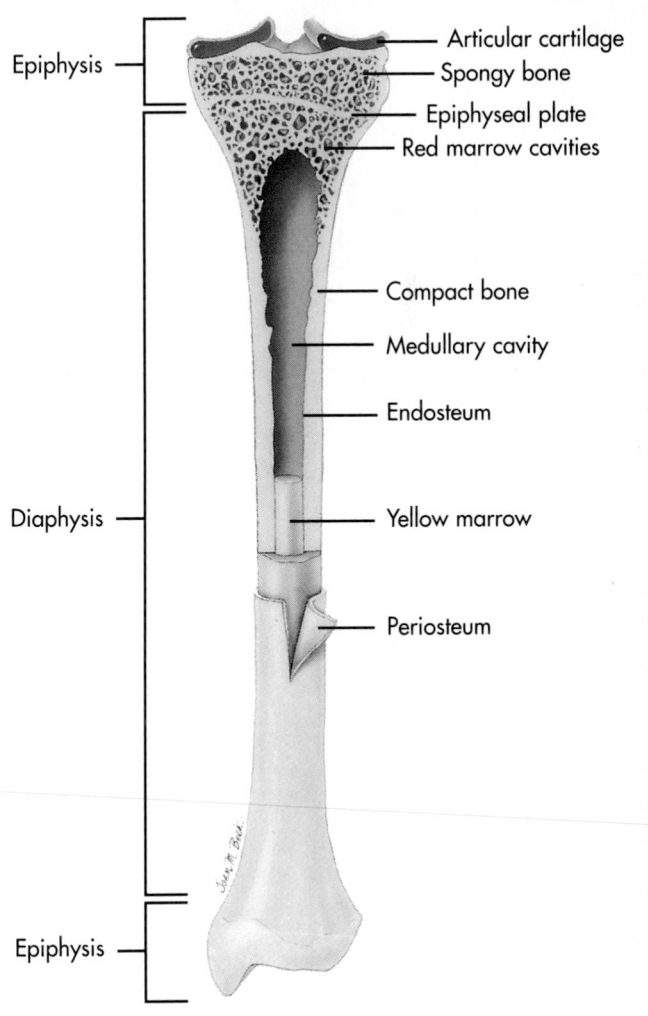

Epiphysis	Articular cartilage
	Spongy bone
	Epiphyseal plate
	Red marrow cavities
	Compact bone
	Medullary cavity
	Endosteum
Diaphysis	Yellow marrow
	Periosteum
Epiphysis	

Longitudinal section of a long bone.

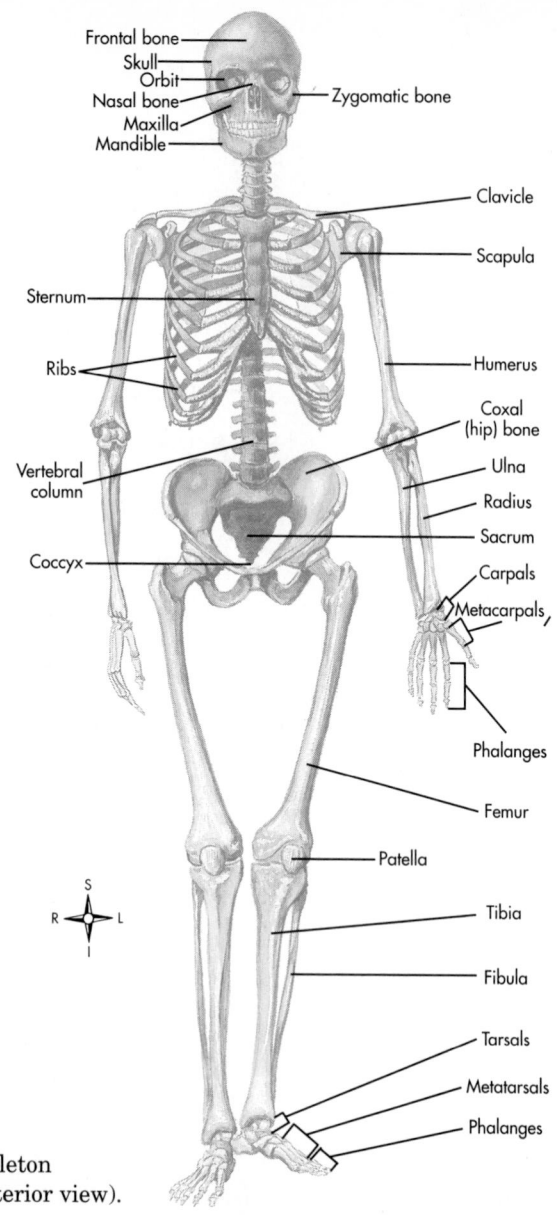

Frontal bone
Skull
Orbit
Nasal bone
Maxilla
Mandible
Zygomatic bone
Clavicle
Scapula
Sternum
Humerus
Ribs
Coxal (hip) bone
Vertebral column
Ulna
Radius
Sacrum
Coccyx
Carpals
Metacarpals
Phalanges
Femur
Patella
Tibia
Fibula
Tarsals
Metatarsals
Phalanges

Skeleton
(anterior view).

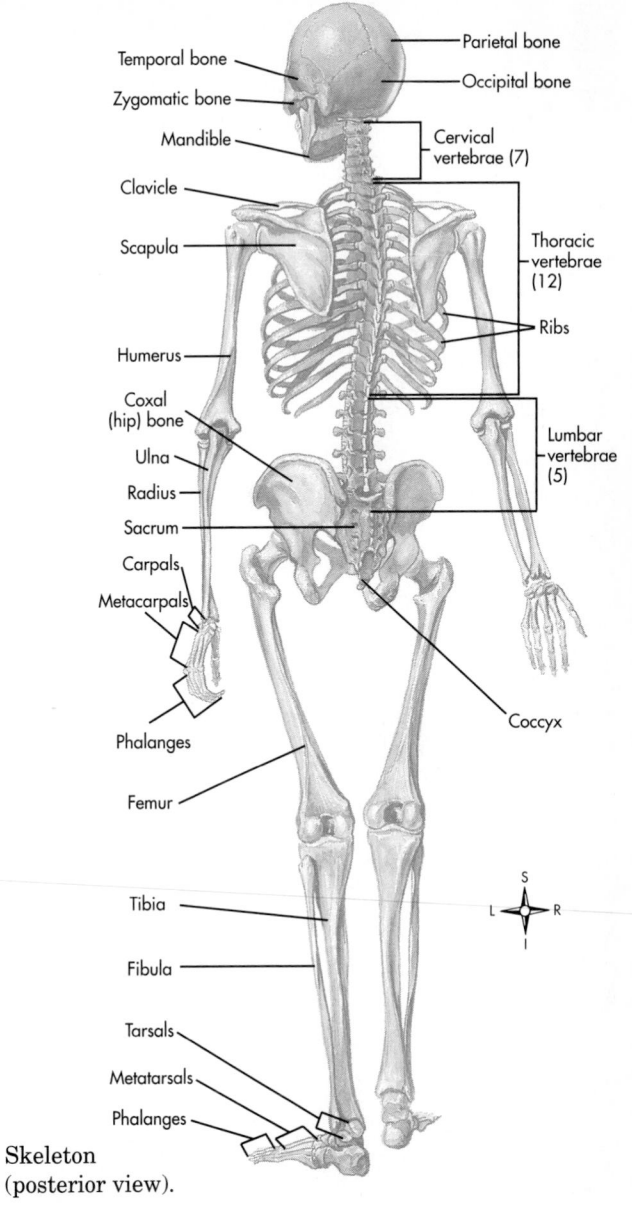

Temporal bone

Zygomatic bone

Mandible

Clavicle

Scapula

Humerus

Coxal (hip) bone

Ulna

Radius

Sacrum

Carpals

Metacarpals

Phalanges

Femur

Tibia

Fibula

Tarsals

Metatarsals

Phalanges

Parietal bone

Occipital bone

Cervical vertebrae (7)

Thoracic vertebrae (12)

Ribs

Lumbar vertebrae (5)

Coccyx

Skeleton (posterior view).

BONES OF SKELETON (206 TOTAL)*

AXIAL SKELETON (80 bones total)

Part of Body	Name of Bone
Skull (28 bones total)	
Cranium (8 bones)	Frontal (1)
	Parietal (2)
	Temporal (2)
	Occipital (1)
	Sphenoid (1)
	Ethmoid (1)
Face (14 bones)	Nasal (2)
	Maxillary (2)
	Zygomatic (molar) (2)
	Mandible (1)
	Lacrimal (2)
	Palatine (2)
	Inferior conchae (turbinates) (2)
	Vomer (1)
Ear bones (6 bones)	Malleus (hammer) (2)
	Incus (anvil) (2)
	Stapes (stirrup) (2)
Hyoid bone (1)	
Spinal column (26 bones total)	Cervical vertebrae (7)
	Thoracic vertebrae (12)
	Lumbar vertebrae (5)
	Sacrum (1)
	Coccyx (1)

AXIAL SKELETON—cont'd

Part of Body	Name of Bone
Sternum and ribs (25 bones total)	Sternum (1)
	True ribs (14)
	False ribs (10)

APPENDICULAR SKELETON (126 bones total)

Part of Body	Name of Bone
Upper extremities (including shoulder girdle) (64 bones total)	Clavicle (2)
	Scapula (2)
	Humerus (2)
	Radius (2)
	Ulna (2)
	Carpals (16)
	Metacarpals (10)
	Phalanges (28)
Lower extremities (including hip girdle) (62 bones total)	Coxal bones (2)
	Femur (2)
	Patella (2)
	Tibia (2)
	Fibula (2)
	Tarsals (14)
	Metatarsals (10)
	Phalanges (28)

TERMS USED TO DESCRIBE BONE MARKINGS

Term	Meaning
Angle	A corner
Body	The main portion of a bone
Condyle	Rounded bump; usually fits into a fossa on another bone, forming a joint
Crest	Moderately raised ridge; generally a site for muscle attachment
Epicondyle	Bump near a condyle; often gives the appearance of a "bump on a bump"; for muscle attachment
Facet	Flat surface that forms a joint with another facet or flat bone
Fissure	Long, cracklike hole for blood vessels and nerves
Foramen	Round hole for vessels and nerves (pl. *foramina*)
Fossa	Depression; often receives an articulating bone (pl. *fossae*)
Head	Distinct epiphysis on a long bone, separated from the shaft by a narrowed portion (or neck)
Line	Similar to a crest but not raised as much (is often rather faint)

Term	Meaning
Margin	Edge of a flat bone or flat portion of edge of a flat area
Meatus	Tubelike opening or channel (pl. *meati*)
Neck	A narrowed portion, usually at the base of a head
Notch	A V-like depression in the margin or edge of a flat area
Process	A raised area or projection
Ramus	Curved portion of a bone, like a ram's horn (pl. *rami*)
Sinus	Cavity within a bone
Spine	Similar to a crest but raised more; a sharp, pointed process; for muscle attachment
Sulcus	Groove or elongated depression (pl. *sulci*)
Trochanter	Large bump for muscle attachment (larger than tubercle or tuberosity)
Tuberosity	Oblong, raised bump, usually for muscle attachment; small tuberosity is called a *tubercle*

11

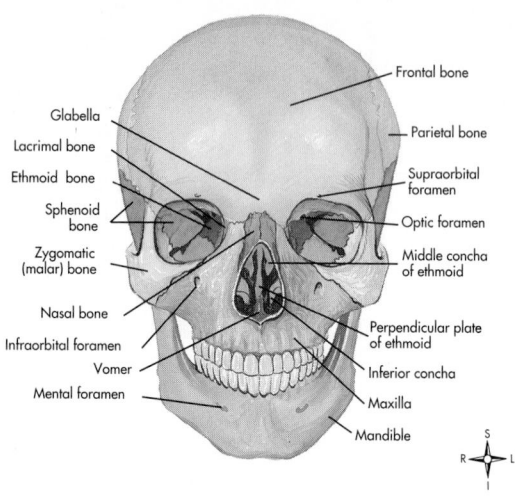

Frontal bone
Parietal bone
Glabella
Lacrimal bone
Ethmoid bone
Sphenoid bone
Supraorbital foramen
Optic foramen
Zygomatic (malar) bone
Middle concha of ethmoid
Nasal bone
Perpendicular plate of ethmoid
Infraorbital foramen
Vomer
Inferior concha
Mental foramen
Maxilla
Mandible

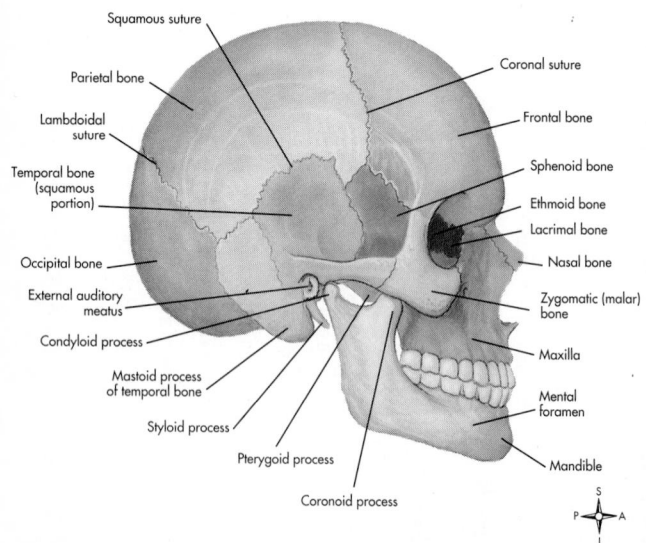

Squamous suture
Coronal suture
Parietal bone
Frontal bone
Lambdoidal suture
Sphenoid bone
Temporal bone (squamous portion)
Ethmoid bone
Lacrimal bone
Nasal bone
Occipital bone
Zygomatic (malar) bone
External auditory meatus
Condyloid process
Maxilla
Mastoid process of temporal bone
Mental foramen
Styloid process
Pterygoid process
Mandible
Coronoid process

CRANIAL BONES AND THEIR MARKINGS

Bones and Markings	Description	Bones and Markings	Description
FRONTAL	Forehead bone; also forms most of roof of orbits (eye sockets) and anterior part of cranial floor	**PARIETAL**	Prominent, bulging bones behind frontal bone; forms top sides of cranial cavity
Supraorbital margin	Arched ridge just below eyebrow, forms upper edge of orbit	**SPHENOID**	Keystone of cranial floor; forms its midportion; resembles bat with wings outstretched and legs extended downward posteriorly; lies behind and slightly above nose and throat; forms part of floor and sidewalls of orbit
Frontal sinuses	Cavities inside bone just above supraorbital margin; lined with mucosa; contain air	**Body**	Hollow, cubelike central portion
Frontal tuberosities	Bulge above each orbit; most prominent part of forehead	**Greater wings**	Lateral projections from body, form part of outer wall of orbit
Superciliary ridges	Ridges caused by projection of frontal sinuses; eyebrows lie superficial to these ridges	**Lesser wings**	Thin, triangular projections from upper part of sphenoid body; form posterior part of roof of orbit
Supraorbital foramen (sometimes notch)	Foramen or notch in supraorbital margin slightly medial to its midpoint; transmits supraorbital nerve and blood vessels	**Sella turcica (or Turk's saddle)**	Saddle-shaped depression on upper surface of sphenoid body; contains pituitary gland
Glabella	Smooth area between superciliary ridges and above nose		

(continued)

13

CRANIAL BONES AND THEIR MARKINGS—cont'd

Bones and Markings	Description
SPHENOID—cont'd	
Sphenoid sinuses	Irregular air-filled mucosa-lined spaces within central part of sphenoid
Pterygoid processes	Downward projections on either side where body and greater wing unite; comparable to extended legs of bat if entire bone is likened to this animal; form part of lateral nasal wall
Optic foramen	Opening into orbit at root of lesser wing; transmits optic nerve
Superior orbital fissure	Slitlike opening into orbit; lateral to optic foramen; transmits third, fourth, and part of fifth cranial nerves
Foramen rotundum	Opening in greater wing that transmits maxillary division of fifth cranial nerve
Foramen ovale	Opening in greater wing that transmits mandibular division of fifth cranial nerve

Bones and Markings	Description
TEMPORAL—cont'd	
External auditory meatus (or canal)	Tube extending into temporal bone from external ear opening to tympanic membrane
Zygomatic process	Projection that articulates with molar (or zygomatic) bone
Internal auditory meatus	Fairly large opening on posterior surface of petrous portion of bone; transmits eighth cranial nerve to inner ear and seventh cranial nerve on its way to facial structures
Mandibular fossa	Oval-shaped depression anterior to external auditory meatus; forms socket for condyle of mandible
Styloid process	Slender spike of bone extending downward and forward from undersurface of bone anterior to mastoid process; often broken off in dry skull; several neck muscles and ligaments attach to styloid process

14

Foramen lacerum	Opening at the junction of the sphenoid, temporal, and occipital bones; transmits branch of the ascending pharyngeal artery
Foramen spinosum	Opening in greater wing that transmits the middle meningeal artery to supply meninges
TEMPORAL	Form lower sides of cranium and part of cranial floor; contain middle and inner ear structures
Squamous portion	Thin, flaring upper part of bone
Mastoid portion	Rough-surfaced lower part of bone posterior to external auditory meatus
Petrous portion	Wedge-shaped process that forms part of center section of cranial floor between sphenoid and occipital bones; name derived from Greek word for stone because of extreme hardness of this process; houses middle and inner ear structures
Mastoid process	Protuberance just behind ear
Mastoid air cells	Air-filled mucosa-lined spaces within mastoid process
Stylomastoid foramen	Opening between styloid and mastoid processes where facial nerve emerges from cranial cavity
Jugular fossa	Depression on undersurface of petrous portion; dilated beginning of internal jugular vein lodged here
Jugular foramen	Opening in suture between petrous portion and occipital bone; transmits lateral sinus and ninth, tenth, and eleventh cranial nerves
Carotid canal (or foramen)	Channel in petrous portion; best seen from undersurface of skull; transmits internal carotid artery
OCCIPITAL	Forms posterior part of cranial floor and walls
Foramen magnum	Hole through which spinal cord enters cranial cavity
Condyles	Convex, oval processes on either side of foramen magnum; articulate with depressions on first cervical vertebra

(continued)

CRANIAL BONES AND THEIR MARKINGS—cont'd

Bones and Markings	Description	Bones and Markings	Description
OCCIPITAL—cont'd		**ETHMOID—cont'd**	
External occipital protuberance	Prominent projection on posterior surface in midline short distance above foramen magnum; can be felt as definite bump	**Horizontal (cribriform) plate**	Olfactory nerves pass through numerous holes in this plate
Superior nuchal line	Curved ridge extending laterally from external occipital protuberance	**Crista galli**	Meninges (membranes around the brain) attach to this process
Inferior nuchal line	Less well-defined ridge paralleling superior nuchal line a short distance below it	**Perpendicular plate**	Forms upper part of nasal septum
Internal occipital protuberance	Projection in midline on inner surface of bone; grooves for lateral sinuses extend laterally from this process and one for sagittal sinus extends upward from it	**Ethmoid sinuses**	Honeycombed, mucosa-lined air spaces within lateral masses of bone
		Superior and middle conchae (turbinates)	Help to form lateral walls of nose
ETHMOID	Complicated irregular bone that helps make up anterior portion of cranial floor, medial wall of orbits, upper parts of nasal septum, and sidewalls and part of nasal roof; lies anterior to sphenoid and posterior to nasal bones	**Lateral masses**	Compose sides of bone; contain many air spaces (ethmoid cells or sinuses); inner surface forms superior and middle conchae

FACIAL BONES AND THEIR MARKINGS

Bones and Markings	Description	Bones and Markings	Description
PALATINE	Form posterior part of hard palate, floor, and part of sidewalls of nasal cavity and floor of orbit	**MANDIBLE**—cont'd	
		Alveolar process	Teeth set into this arch
Horizontal plate	Joined to palatine processes of maxillae to complete part of hard palate	**Mandibular foramen**	Opening on inner surface of ramus; transmits nerves and vessels to lower teeth
MANDIBLE	Lower jawbone; largest, strongest bone of face	**Mental foramen**	Opening on outer surface below space between two bicuspids; transmits terminal branches of nerves and vessels that enter bone through mandibular foramen; dentists inject anesthetics through these foramina
Body	Main part of bone; forms chin		
Ramus	Process, one on either side, that projects upward from posterior part of body	**Coronoid process**	Projection upward from anterior part of each ramus; temporal muscle inserts here
Condyle (or head)	Part of each ramus that articulates with mandibular fossa of temporal bone	**Angle**	Juncture of posterior and inferior margins of ramus
Neck	Constricted part just below condyles		

(continued)

17

Bones and Markings	Description
NASAL	Small bones forming upper part of bridge of nose
ZYGOMATIC	Cheekbones; form part of floor and sidewall of orbit
LACRIMAL	Thin bones about size and shape of fingernail; posterior and lateral to nasal bones in medial wall of orbit; help form sidewall of nasal cavity, often missing in dry skull
INFERIOR NASAL CONCHAE (turbinates)	Thin scroll of bone forming kind of shelf along inner surface of sidewall of nasal cavity; lies above roof of mouth
VOMER	Forms lower and posterior part of nasal septum; shaped like ploughshare

Bones and Markings	Description
MAXILLA	Upper jaw bones; form part of floor of orbit, anterior part of roof of mouth, and floor of nose and part of sidewalls of nose
Alveolar process	Arch containing teeth
Maxillary sinus (antrum of Highmore)	Large air-filled mucosa-lined cavity within body of each maxilla; largest of sinuses
Palatine process	Horizontal inward projection from alveolar process; forms anterior and larger part of hard palate
Infraorbital foramen	Hole on external surface just below orbit; transmits vessels and nerves
Lacrimal groove	Groove on inner surface; joined by similar groove on lacrimal bone to form canal housing nasolacrimal duct

SPECIAL FEATURES OF SKULL

Feature	Description	Feature	Description
SUTURES		**FONTANELS—cont'd**	
Squamous	Immovable joints between skull bones	**Frontal (or anterior)**	At intersection of sagittal and coronal sutures (juncture of parietal bones and frontal bone); diamond shaped; largest of fontanels; usually closed by 1½ years of age
	Line of articulation along top curved edge of temporal bone		
Coronal	Joint between parietal bones and frontal bone		
Lambdoidal	Joint between parietal bones and occipital bone	**Occipital (or posterior)**	At intersection of sagittal and lambdoidal sutures (juncture of parietal bones and occipital bone); triangular in shape; usually closed by second month
Sagittal	Joint between right and left parietal bones		
FONTANELS	"Soft spots" where ossification is incomplete at birth; allow some compression of skull during birth; also important in determining position of head before delivery; six such areas located at angles of parietal bones	**Sphenoid (or anterolateral)**	At juncture of frontal, parietal, temporal, and sphenoid bones
		Mastoid (or posterolateral)	At juncture of parietal, occipital, and temporal bones; usually closed by second year

(continued)

SPECIAL FEATURES OF SKULL—cont'd

Feature	Description	Feature	Description
AIR SINUSES	Spaces or cavities within bones; those that communicate with nose called *paranasal sinuses* (frontal, sphenoidal, ethmoidal, and maxillary); mastoid cells communicate with middle ear rather than nose, therefore not included among paranasal sinuses	**ORBITS FORMED BY—cont'd**	
		Palatine	Floor
		NASAL SEPTUM FORMED BY	Partition in midline of nasal cavity; separates cavity into right and left halves
		Perpendicular plate of ethmoid bone	Forms upper part of septum
ORBITS FORMED BY		Vomer bone	Forms lower, posterior part
Frontal	Roof of orbit	Cartilage	Forms anterior part
Ethmoid	Medial wall	**WORMIAN BONES**	Small islets of bones with suture
Lacrimal	Medial wall	**MALLEUS, INCUS, STAPLES**	Tiny bones, referred to as auditory ossicles, in middle ear cavity in temporal bones; resemble, respectively, miniature hammer, anvil, and stirrup
Sphenoid	Lateral wall		
Zygomatic	Lateral wall		
Maxillary	Floor		

HYOID, VERTEBRAE, AND THORACIC BONES AND THEIR MARKINGS

Bones and Markings	Description	Bones and Markings	Description
HYOID	U-shaped bone in neck between mandible and upper part of larynx; distinctive as only bone in body not forming a joint with any other bone; suspended by ligaments from styloid processes of temporal bones	**VERTEBRAL COLUMN—cont'd**	
		Body	Main part; flat, round mass located anteriorly; supporting or weight-bearing part of vertebra
		Pedicles	Short projections extending posteriorly from body
VERTEBRAL COLUMN	Not actually a column but a flexible, segmented curved rod; forms axis of body; head balanced above, ribs and viscera suspended in front, and lower extremities attached below; encloses spinal cord	**Lamina**	Posterior part of vertebra to which pedicles join and from which processes project
		Neural arch	Formed by pedicles and laminae; protects spinal cord posteriorly; congenital absence of one or more neural arches is known as *spina bifida* (cord may protrude right through skin)
General features	Anterior part of each vertebra (except first two cervical) consists of body; posterior part of vertebrae consists of neural arch, which, in turn, consists of two pedicles, two laminae, and seven processes projecting from laminae	**Spinous process**	Sharp process projecting inferiorly from laminae in midline

(continued)

21

HYOID, VERTEBRAE, AND THORACIC BONES AND THEIR MARKINGS—cont'd

Bones and Markings	Description	Bones and Markings	Description
VERTEBRAL COLUMN—cont'd		**THORACIC VERTEBRAE—cont'd**	
Transverse processes	Right and left lateral projections from laminae		massive bodies than cervical vertebrae; no transverse foramina; two sets of facets for articulations with corresponding rib: one on body, second on transverse process; upper thoracic vertebrae with elongated spinous process
Superior articulating processes	Project upward from laminae		
Inferior articulating processes	Project downward from laminae; articulate with superior articulating processes of vertebra below	**LUMBAR VERTEBRAE**	Next 5 vertebrae; strong, massive; superior articulating processes directed medially instead of upward; inferior articulating processes, laterally instead of downward; short, blunt spinous process
Spinal foramen	Hole in center of vertebra formed by union of body, pedicles, and laminae; spinal foramina, when vertebrae, superimposed one on other, form spinal cavity that houses spinal cord		
Intervertebral foramina	Opening between vertebrae though which spinal nerves emerge	**SACRUM**	Five separate vertebrae until about 25 years of age; then fused to form one wedge-shaped bone
CERVICAL VERTEBRAE	First or upper seven vertebrae; foramen in each transverse process for transmis-	Sacral promontory	Protuberance from anterior, upper border of sacrum into pelvis; of obstetrical

sion of vertebral artery, vein, and plexus of nerves; short bifurcated spinous processes except on seventh vertebra, where it is extra long and may be felt as protrusion when head bent forward; bodies of these vertebrae small, whereas spinal foramina large and triangular

Atlas
First cervical vertebra; lacks body and spinous process; superior articulating processes concave ovals that act as rockerlike cradles for condyles of occipital bone; named *atlas* because it supports the head as Atlas supports the world in Greek mythology

Axis (epistropheus)
Second cervical vertebra, so named because atlas rotates about this bone in rotating movements of head; *dens*, or odontoid process, peglike projection upward from body of axis, forming pivot for rotation of atlas

THORACIC VERTEBRAE
Next 12 vertebrae; 12 pairs of ribs attached to these; stronger, with more

importance because its size limits anteroposterior diameter of pelvic inlet

Coccyx
Four or five separate vertebrae in child but fused into one in adult

CURVES
Curves have great structural importance because they increase carrying strength of vertebral column, make balance possible in upright position (if column were straight, weight of viscera would pull body forward), absorb jolts from walking (straight column would transmit jolts straight to head), and protect column from fracture

Primary
Column curves at birth from head to sacrum with convexity posteriorly; after child stands, convexity persists only in *thoracic* and *sacral* regions, which therefore are called *primary curves*

Secondary
Concavities in *cervical* and *lumbar* regions; cervical concavity results from

(continued)

Bones and Markings	Description	Bones and Markings	Description
CURVES—cont'd	infant's attempts to hold head erect (2 to 4 months); lumbar concavity, from balancing efforts in learning to walk (10 to 18 months)	**False ribs (cont'd)**	attach by means of costal cartilage of seventh ribs; last two pairs do not attach to sternum at all, therefore called "floating" ribs
STERNUM	Breastbone; flat dagger-shaped bone; sternum, ribs, and thoracic vertebrae together form bony cage known as *thorax*	**Head**	Projection at posterior end of rib; articulates with corresponding thoracic vertebra and one above, except last three pairs, which join corresponding vertebrae only
Body	Main central part of bone	**Neck**	Constricted portion just below head
Manubrium	Flaring, upper part	**Tubercle**	Small knob just below neck; articulates with transverse process of corresponding thoracic vertebra; missing in lowest 3 ribs
Xiphoid process	Projection of cartilage at lower border of bone		
RIBS		**Body or shaft**	Main part of rib
True ribs	Upper seven pairs; fasten to sternum by costal cartilages	**Costal cartilage**	Cartilage at sternal end of true ribs; attaches ribs (except floating ribs) to sternum
False ribs	False ribs do not attach to sternum directly; upper three pairs of false ribs		

UPPER EXTREMITY BONES AND THEIR MARKINGS

Bones and Markings	Description	Bones and Markings	Description
CLAVICLE	Collar bones; shoulder girdle joined to axial skeleton by articulation of clavicles with sternum (scapula does not form joint with axial skeleton)	**SCAPULA—cont'd**	
		Coracoid process	Projection on anterior surface from upper border of bone; may be felt in groove between deltoid and pectoralis major muscles, about 1 inch below clavicle
SCAPULA	Shoulder blades; scapulae and clavicles together comprise shoulder girdle	**Glenoid cavity**	Arm socket
Borders		**HUMERUS**	Long bone of upper arm
Superior	Upper margin	**Head**	Smooth, hemispherical enlargement at proximal end of humerus
Vertebral	Margin toward vertebral column		
Axillary	Lateral margin	**Anatomical neck**	Oblique groove just below head
Spine	Sharp ridge running diagonally across posterior surface of shoulder blade	**Greater tubercle**	Rounded projection lateral to head on anterior surface
Acromion process	Slightly flaring projection at lateral end of scapular spine; may be felt as tip of shoulder; articulates with clavicle	**Lesser tubercle**	Prominent projection on anterior surface just below anatomical neck

(continued)

25

UPPER EXTREMITY BONES AND THEIR MARKINGS—cont'd

Bones and Markings	Description	Bones and Markings	Description
HUMERUS—cont'd		**ULNA**	Bone of little finger side of forearm; longer than radius
Intertubercular groove	Deep groove between greater and lesser tubercles; long tendon of biceps muscle lodges here	**Olecranon process**	Elbow
Surgical neck	Region just below tubercles; so named because of its liability to fracture	**Coronoid process**	Projection on anterior surface of proximal end of ulna; trochlea of humerus fits snugly between olecranon and coronoid processes
Deltoid tuberosity	V-shaped, rough area about midway down shaft where deltoid muscle inserts	**Semilunar notch**	Curved notch between olecranon and coronoid process into which trochlea fits
Radial groove	Groove running obliquely downward from deltoid tuberosity; lodges radial nerve	**Radial notch**	Curved notch lateral and inferior to semilunar notch; head of radius fits into this concavity
Epicondyles (medial and lateral)	Rough projections at both sides of distal end	**Head**	Rounded process at distal end; does not articulate with wrist bones but with fibrocartilaginous disk
Capitulum	Rounded knob below lateral epicondyle; articulates with radius; sometimes called *radial head* of humerus		

Trochlea	Projection with deep depression through center similar to shape of pulley; articulates with ulna
Olecranon fossa	Depression on posterior surface just above trochlea; receives olecranon process of ulna when lower arm extends
Coronoid fossa	Depression on anterior surface above trochlea; receives coronoid process of ulna in flexion of lower arm
RADIUS	
Head	Bone of thumb side of forearm
	Disk-shaped process forming proximal end of radius; articulates with capitulum of humerus and with radial notch of ulna
Radial tuberosity	Roughened projection on ulnar side, short distance below head; biceps muscle inserts here
Styloid process	Protuberance at distal end on lateral surface (with forearm in anatomical position)
Styloid process	Sharp protuberance at distal end; can be seen from outside on posterior surface
CARPALS	Wrist bones; arranged in two rows at proximal end of hand; proximal row (from little finger toward thumb)—*pisiform, triquetrum, lunate,* and *scaphoid;* distal row—*hamate, capitate, trapezoid,* and *trapezium*
METACARPALS	Long bones forming framework of palm of hand; numbered I through V
PHALANGES	Miniature long bones of fingers, three (proximal, middle, distal) in each finger, two (proximal, distal) in each thumb

LOWER EXTREMITY BONES AND THEIR MARKINGS

Bones and Markings	Description	Bones and Markings	Description
COXAL	Large hip bone; with sacrum and coccyx, forms basinlike pelvic cavity; lower extremities attached to axial skeleton by coxal bones	**COXAL—cont'd**	
		Obturator foramen	Large hole in anterior surface of os coxa; formed by pubis and ischium; largest foramen in body
Ilium	Upper, flaring portion	Pelvic brim (or inlet)	Boundary of aperture leading into true pelvis; formed by pubic crests, iliopectineal lines, and sacral promontory; size and shape of this inlet have obstetrical importance, since if any of its diameters too small, infant skull cannot enter true pelvis for natural birth
Ischium	Lower, posterior portion		
Pubic bone (pubis)	Medial, anterior section		
Acetabulum	Hip socket; formed by union of ilium, ischium, and pubis		
Iliac crests	Upper, curving boundary of ilium	True pelvis (or pelvis minor)	Space below pelvic brim; true "basin" with bone and muscle walls and muscle floor; pelvic organs located in this space
Iliac spines			
Anterior superior	Prominent projection at anterior end of iliac crest; can be felt externally as "point" of hip	False pelvis (or pelvis major)	Broad, shallow space above pelvic brim, or pelvic inlet; name "false pelvis" is misleading, since this space is actually
Anterior inferior	Less prominent projection short distance below anterior superior spine		

Posterior superior	At posterior end of iliac crest
Posterior inferior	Just below posterior superior spine
Greater sciatic notch	Large notch on posterior surface of ilium just below posterior inferior spine
Ischial tuberosity	Large, rough, quadrilateral process forming inferior part of ischium; in erect sitting position body rests on these tuberosities
Ischial spine	Pointed projection just above tuberosity
Symphysis pubis	Cartilaginous, amphiarthrotic joint between pubic bones
Superior ramus of pubis	Part of pubis lying between symphysis and acetabulum; forms upper part of obturator foramen
Inferior ramus	Part extending down from symphysis; unites with ischium
Pubic arch	Angle formed by two inferior rami
Pubic crest	Upper margin of superior ramus
Pubic tubercle	Rounded process at end of crest
	part of abdominal cavity, not pelvic cavity
Pelvic outlet	Irregular circumference marking lower limits of true pelvis; bounded by tip of coccyx and two ischial tuberosities
Pelvic girdle (or bony pelvis)	Complete bony ring; composed of two hip bones (ossa coxae), sacrum, and coccyx; forms firm base by which trunk rests on thighs and for attachment of lower extremities to axial skeleton
FEMUR	Thigh bone; largest, strongest bone of body
Head	Rounded, upper end of bone; fits into acetabulum
Neck	Constricted portion just below head
Greater trochanter	Protuberance located inferiorly and laterally to head
Lesser trochanter	Small protuberance located inferiorly and medially to greater trochanter

(continued)

LOWER EXTREMITY BONES AND THEIR MARKINGS—cont'd

Bones and Markings	Description	Bones and Markings	Description
FEMUR—cont'd		**TIBIA—cont'd**	
Intertrochanteric line	Line extending between greater and lesser trochanter	Crest	Sharp ridge on anterior surface
		Tibial tuberosity	Projection in midline on anterior surface
Linea aspera	Prominent ridge extending lengthwise along concave posterior surface	Medial malleolus	Rounded downward projection at distal end of tibia; forms prominence on medial surface of ankle
Supracondylar ridges	Two ridges formed by division of linea aspera at its lower end; medial supracondylar ridge extends inward to inner condyle, lateral ridge to outer condyle	FIBULA	Long, slender bone of lateral side of lower leg
Condyles	Large, rounded bulges at distal end of femur; one medial and one lateral	Lateral malleolus	Rounded prominence at distal end of fibula; forms prominence on lateral surface of ankle
Epicondyles	Blunt projections from the sides of the condyles; one on the medial aspect and one on the lateral aspect	TARSALS	Bones that form heel and proximal or posterior half of foot
		Calcaneus	Heel bone

Talus	Uppermost of tarsals; articulates with tibia and fibula; boxed in by medial and lateral malleoli
Longitudinal arches	Tarsals and metatarsals so arranged as to form arch from front to back of foot
Medial	Formed by calcaneus, talus, navicular, cuneiforms, and three medial metatarsals
Lateral	Formed by calcaneus, cuboid, and two lateral metatarsals
Transverse (or metatarsal) arch	Metatarsals and distal row of tarsals (cuneiforms and cuboid) so articulated as to form arch across foot; bones kept in two arched positions by means of powerful ligaments in sole of foot and by muscles and tendons
METATARSALS	Long bones of feet
PHALANGES	Miniature long bones of toes; two in each great toe; three in other toes

Adductor tubercle	Small projection just above medial condyle; marks termination of medial supracondylar ridge
Trochlea	Smooth depression between condyles on anterior surface; articulates with patella
Intercondyloid fossa (notch)	Deep depression between condyles on posterior surface; cruciate ligaments that help bind femur to tibia lodge in this notch
PATELLA	Kneecap; largest sesamoid bone of body; embedded in tendon of quadriceps femoris muscle
TIBIA	Shin bone
Condyles	Bulging prominences at proximal end of tibia; upper surfaces concave for articulation with femur
Intercondylar eminence	Upward projection on articular surface between condyles

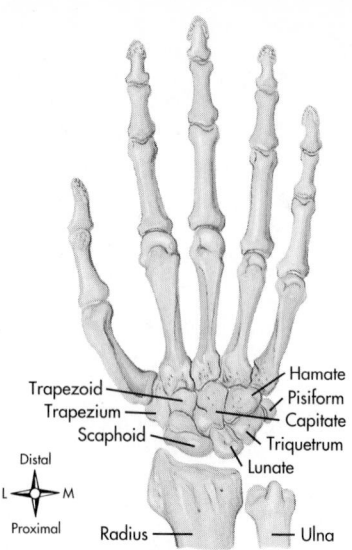

Right hand and wrist: Dorsal view.

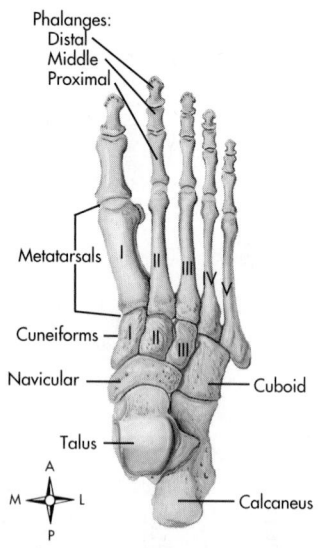

Right foot viewed from above. Tarsal bones consist of
cuneiforms, navicular, talus, cuboid, and calcaneus.

CLASSIFICATION OF JOINTS

Types	Examples	Structural Features	Movements
FIBROUS JOINTS			
Syndesmoses	Joints between distal ends of radius and ulna	Fibrous bands (ligaments) connect articulating bones	Slight
Sutures	Joints between skull bones	Teeth-like projections of articulating bones interlock with thin layer of fibrous tissue connecting them	None
Gomphoses	Joints between roots of teeth and jaw bones	Fibrous tissue connects roots of teeth to alveolar processes	None
CARTILAGINOUS JOINTS			
Synchondroses	Costal cartilage attachments of first rib to sternum; epiphyseal plate between diaphysis and epiphysis of growing long bone	Hyaline cartilage connects articulating bones	Slight
Symphyses	Symphysis pubis; joints between *bodies* of vertebrae	Fibrocartilage between articulating bones	Slight

CLASSIFICATION OF SYNOVIAL JOINTS

Types	Examples	Structural Features	Movements
UNIAXIAL			Around one axis; in one plane
Hinge	Elbow joint	Spool-shaped process fits into concave socket	Flexion and extension only
Pivot	Joint between first and second cervical vertebrae	Arch-shaped process fits around peglike process	Rotation
BIAXIAL			Around two axes, perpendicular to each other; in two planes
Saddle	Thumb joint between first metacarpal and carpal bone	Saddle-shaped bone fits into socket that is concave-convex-concave	Flexion, extension in one plane; abduction; adduction in other plane; opposing thumb to fingers

Condyloid (ellipsoidal)	 Oval condyle fits into elliptical socket	Joint between radius and carpal bones	Flexion, extension in one plane; abduction, adduction in other plane
MULTIAXIAL			Around many axes
Ball and socket	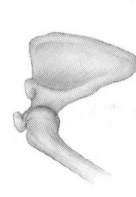 Ball-shaped process fits into concave socket	Shoulder joint and hip joint	Widest range of movements; flexion, extension, abduction, adduction, rotation, circumduction
Gliding	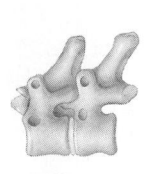 Relatively flat articulating surfaces	Joints between articular facets of adjacent vertebrae; joints between carpal and tarsal bones	Gliding movements without any angular or circular movements

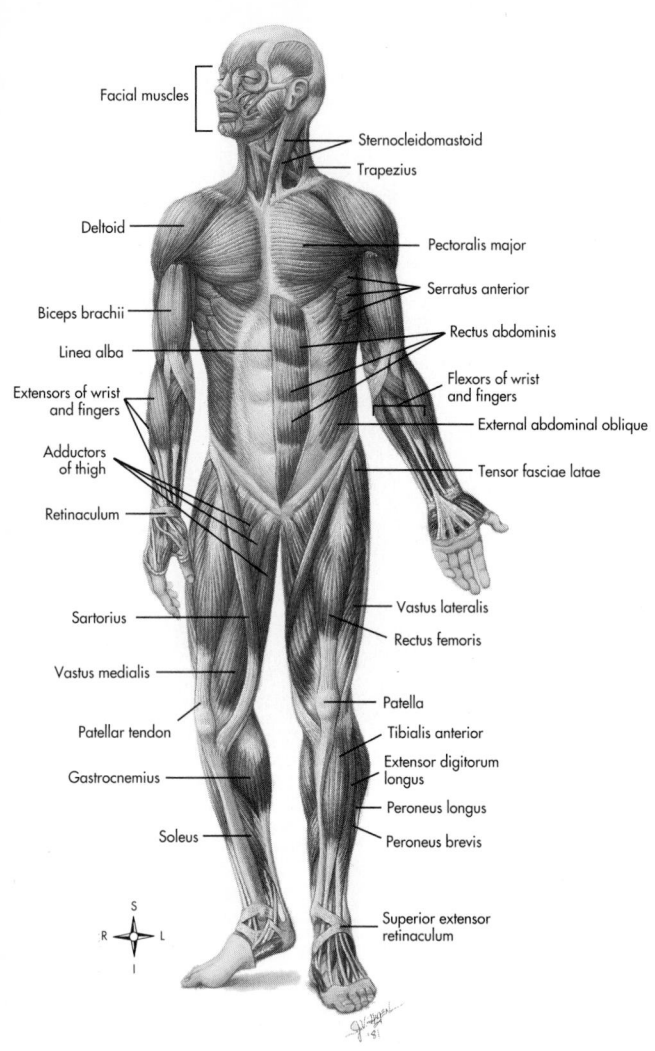

Facial muscles

Sternocleidomastoid

Trapezius

Deltoid

Pectoralis major

Serratus anterior

Biceps brachii

Rectus abdominis

Linea alba

Flexors of wrist
and fingers

Extensors of wrist
and fingers

External abdominal oblique

Adductors
of thigh

Tensor fasciae latae

Retinaculum

Vastus lateralis

Sartorius

Rectus femoris

Vastus medialis

Patella

Patellar tendon

Tibialis anterior

Gastrocnemius

Extensor digitorum
longus

Peroneus longus

Peroneus brevis

Soleus

Superior extensor
retinaculum

S
R — L
I

Muscles (anterior view).

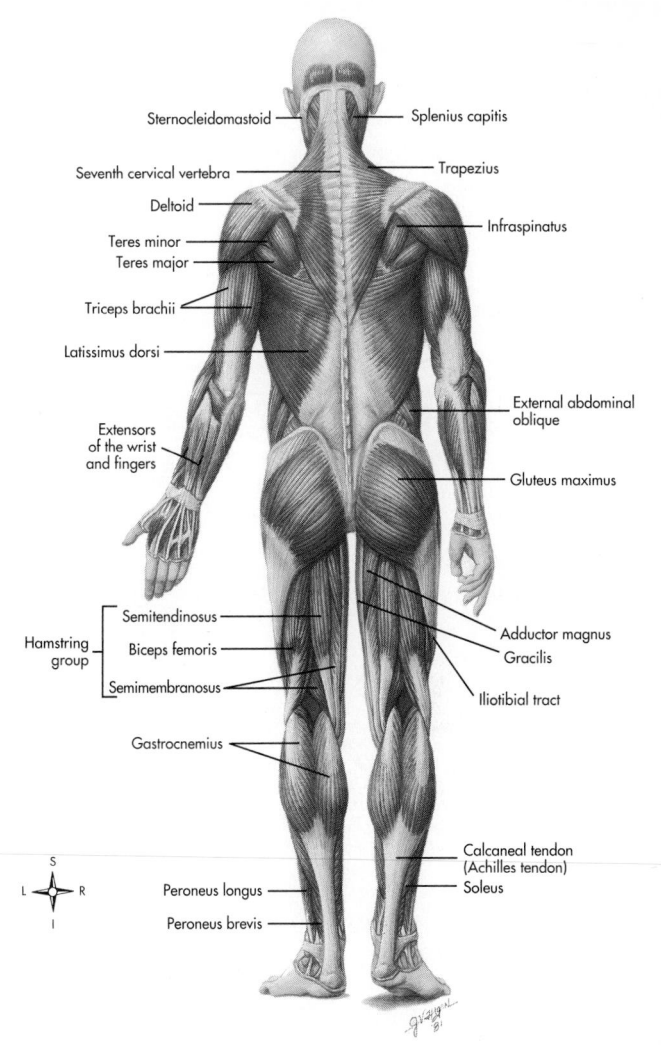

Muscles (posterior view).

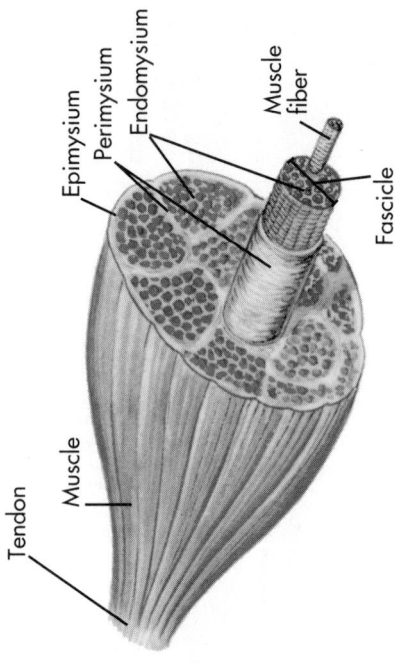

Structure of a skeletal muscle organ. Note that the connective tissue coverings, the epimysium, perimysium, and endomysium, are continuous with each other and with the tendon. Note also that muscle fibers are held together in groups called fascicles by the perimysium.

CHARACTERISTICS OF MUSCLE TISSUES

	Skeletal	Cardiac	Smooth
PRINCIPLE LOCATION	Skeletal muscle organs	Wall of heart	Walls of many hollow organs
PRINCIPLE FUNCTIONS	Movement of bones, heat production, posture	Pumping of blood	Movement in walls of hollow organs (peristalsis, mixing)
TYPE OF CONTROL	Voluntary	Involuntary	Involuntary
STRUCTURAL FEATURES			
Striations	Present	Present	Absent
Nucleus	Many, near sarcolemma	Single	Single, near center of cell
T tubules	Narrow, form triads with SR	Large diameter, form diads with SR, regulate Ca^{++} entry into sarcoplasm	Absent
Sarcoplasmic reticulum	Extensive, stores and releases Ca^{++}	Less extensive than in skeletal muscle	Very poorly developed
Cell junctions	No gap junctions	Intercalated disks	Visceral: many gap junctions Multiunit: few gap junctions
CONTRACTION STYLE	Rapid twitch contractions of motor units usually summate to produce sustained tetanic contractions; must be stimulated by a neuron	Syncytium of fibers compress heart chambers in slow, separate contractions (does not exhibit tetanus or fatigue); exhibits auto-rhythmicity	Visceral: electrically coupled sheets of fibers contract autorhythmically, producing peristalsis or mixing movements Multiunit: individual fibers contract when stimulated by neuron

39

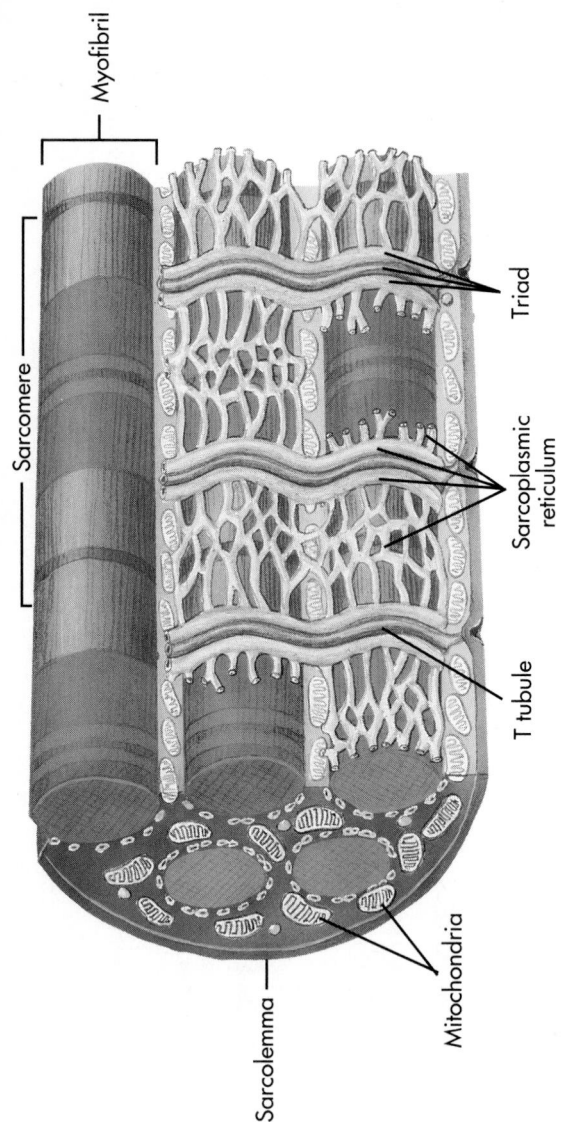

Myofibril

Sarcomere

Triad

Sarcoplasmic
reticulum

T tubule

Sarcolemma

Mitochondria

Skeletal muscle cell.

40

MAJOR EVENTS OF MUSCLE CONTRACTION AND RELAXATION

EXCITATION AND CONTRACTION

1   A nerve impulse reaches the end of a motor neuron, triggering the release of the neurotransmitter *acetylcholine.*

2   Acetylcholine diffuses rapidly across the gap of the neuromuscular junction and binds to acetylcholine receptors on the motor endplate of the muscle fiber.

3   Stimulation of acetylcholine receptors initiates an impulse that travels along the sarcolemma, through the T tubules, to sacs of the sarcoplasmic reticulum (SR).

4   Ca^{++} is released from the SR into the sarcoplasm, where it binds to troponin molecules in the thin myofilaments.

5   Tropomyosin molecules in the thin myofilaments shift, exposing actin's active sites.

6   Energized myosin cross bridges of the thick myofilaments bind to actin and use their energy to pull the thin myofilaments toward the center of each sarcomere. This cycle repeats itself many times per second, as long as ATP is available.

7   As the thin filaments slide past the thick myofilaments, the entire muscle fiber shortens.

RELAXATION

1   After the impulse is over, the SR begins actively pumping Ca^{++} back into its sacs.

2   As Ca^{++} is stripped from troponin molecules in the thin myofilaments, tropomyosin returns to its position, blocking actin's active sites.

3   Myosin cross bridges are prevented from binding to actin and thus can no longer sustain the contraction.

4   Since the thick and thin myofilaments are no longer connected, the muscle fiber may return to its longer, resting length.

41

MUSCLES OF FACIAL EXPRESSION AND OF MASTICATION

Muscle	Origin	Insertion	Function	Nerve Supply
MUSCLES OF FACIAL EXPRESSION				
Occipitofrontalis (epicranius)	Occipital bone	Tissues of eyebrows	Raises eyebrows, wrinkles forehead horizontally	Cranial nerve VII
Corrugator supercilii	Frontal bone (superciliary ridge)	Skin of eyebrow	Wrinkles forehead vertically	Cranial nerve VII
Orbicularis oculi	Encircles eyelid		Closes eye	Cranial nerve VII
Zygomaticus major	Zygomatic bone	Angle of mouth	Laughing (elevates angle of mouth)	Cranial nerve VII
Orbicularis oris	Encircles mouth		Draws lips together	Cranial nerve VII
Buccinator	Maxillae	Skin of sides of mouth	Permits smiling Blowing, as in playing a trumpet	Cranial nerve VII
MUSCLES OF MASTICATION				
Masseter	Zygomatic arch	Mandible (external surface)	Closes jaw	Cranial nerve V
Temporalis	Temporal bone	Mandible	Closes jaw	Cranial nerve V
Pterygoids (lateral and medial)	Undersurface of skull	Mandible (medial surface)	Grates teeth	Cranial nerve V

MUSCLES THAT MOVE THE HEAD

Muscle	Origin	Insertion	Function	Nerve Supply
Sternocleidomastoid	Sternum Clavicle	Temporal bone (mastoid process)	Flexes head (prayer muscle) One muscle alone, rotates head toward opposite side; spasm of this muscle alone or associated with trapezius called *torticollis* or *wryneck*	Accessory nerve
Semispinalis capitis	Vertebrae (transverse processes of upper six thoracic, articular processes of lower four cervical)	Occipital bone (between superior and inferior nuchal lines)	Extends head; bends it laterally	First five cervical nerves
Splenius capitis	Ligamentum nuchae Vertebrae (spinous processes of upper three or four thoracic)	Temporal bone (mastoid process) Occipital bone	Extends head Bends and rotates head toward same side as contracting muscle	Second, third, and fourth cervical nerves

(continued)

43

MUSCLES THAT MOVE THE HEAD—cont'd

Muscle	Origin	Insertion	Function	Nerve Supply
Longissimus capitis	Vertebrae (transverse processes of upper six thoracic, articular processes of lower four cervical)	Temporal bone (mastoid process)	Extends head Bends and rotates head toward contracting side	Multiple innervation

MUSCLES OF THE THORAX

Muscle	Origin	Insertion	Function	Nerve Supply
External intercostals	Rib (lower border; forward fibers)	Rib (upper border of rib below origin)	Elevate ribs	Intercostal nerves
Internal intercostals	Rib (inner surface, lower border, backward fibers)	Rib (upper border of rib below origin)	Depress ribs	Intercostal nerves
Diaphragm	Lower circumference of thorax (of rib cage)	Central tendon of diaphragm	Enlarges thorax, causing inspiration	Phrenic nerves

MUSCLES OF THE ABDOMINAL WALL

Muscle	Origin	Insertion	Function	Nerve Supply
External oblique	Ribs (lower eight)	Ossa coxae (iliac crest and pubis by way of inguinal ligament) Linea alba by way of an aponeurosis	Compresses abdomen Rotates trunk laterally Important postural function of all abdominal muscles is to pull front of pelvis upward, thereby flattening lumbar curve of spine; when these muscles lose their tone, common figure faults of protruding abdomen and lordosis develop	Lower seven intercostal nerves and iliohypogastric nerves
Internal oblique	Ossa coxae (iliac crest and inguinal ligament) Lumbodorsal fascia	Ribs (lower three) Linea alba	Same as external oblique	Last three intercostal nerves; iliohypogastric and ilioinguinal nerves
Transversus abdominis	Ribs (lower six) Ossa coxae (iliac crest, inguinal ligament) Lumbodorsal fascia	Pubic bone Linea alba	Same as external oblique	Last five intercostal nerves; iliohypogastric and ilioinguinal nerves

(continued)

45

MUSCLES OF THE ABDOMINAL WALL—cont'd

Muscle	Origin	Insertion	Function	Nerve Supply
Rectus abdominis	Ossa coxae (pubic bone and symphysis pubis)	Ribs (costal cartilage of fifth, sixth, and seventh ribs) Sternum (xiphoid process)	Same as external oblique; because abdominal muscles compress abdominal cavity, they aid in straining, defecation, forced expiration, childbirth, etc.; abdominal muscles are antagonists of diaphragm, relaxing as it contracts and vice versa Flexes trunk	Last six intercostal nerves

MUSCLES OF THE PELVIC FLOOR

Muscle	Origin	Insertion	Function	Nerve Supply
Levator ani	Pubis and spine of ischium	Coccyx	Together with coccygeus muscles form floor of pelvic cavity and support pelvic organs	Pudendal nerve
Ischiocavernosus	Ischium	Penis or clitoris	Compress base of penis or clitoris	Perineal nerve
Bulbospongiosus Male Female	Bulb of penis Perineum	Perineum and bulb of penis Base of clitoris	Constricts urethra and erects penis Erects clitoris	Pudendal nerve Pudendal nerve
Deep transverse perinei	Ischium	Central tendon (median raphe)	Support pelvic floor	Pudendal nerve
Sphincter urethrae	Pubic ramus	Central tendon (median raphe)	Constrict urethra	Pudendal nerve
Sphincter externus anii	Coccyx	Central tendon (median raphe)	Close anal canal	Pudendal and S4

46

MUSCLES ACTING ON THE SHOULDER GIRDLE

Muscle	Origin	Insertion	Function	Nerve Supply
Trapezius	Occipital bone (protuberance)	Clavicle	Raises or lowers shoulders and shrugs them	Spinal accessory; second, third, and fourth cervical nerves
	Vertebrae (cervical and thoracic)	Scapula (spine and acromion)	Extends head when occiput acts as insertion	
Pectoralis minor	Ribs (second to fifth)	Scapula (coracoid)	Pulls shoulder down and forward	Medial and lateral anterior thoracic nerves
Serratus anterior	Ribs (upper eight or nine)	Scapula (anterior surface, vertebral border)	Pulls shoulder down and forward; abducts and rotates it upward	Long thoracic nerve
Levator scapulae	C1-C4 (transverse processes)	Scapula (superior angle)	Elevates and retracts scapula and abducts neck	Dorsal scapular nerve
Rhomboideus				
Major	T1-T4	Scapula (medial border)	Retracts, rotates, fixes scapula	Dorsal scapular nerve
Minor	C6-C7	Scapula (medial border)	Retracts, rotates, elevates, and fixes scapula	Dorsal scapular nerve

47

MUSCLES THAT MOVE THE UPPER ARM

Muscle	Origin	Insertion	Function	Nerve Supply
AXIAL*				
Pectoralis major	Clavicle (medial half) Sternum Costal cartilages of true ribs	Humerus (greater tubercle)	Flexes upper arm Adducts upper arm anteriorly; draws it across chest	Medial and lateral anterior thoracic nerves
Latissimus dorsi	Vertebrae (spines of lower thoracic, lumbar, and sacral) Ilium (crest) Lumbodorsal fascia	Humerus (intertubercular groove)	Extends upper arm Adducts upper arm posteriorly	Thoracodorsal nerve
SCAPULAR*				
Deltoid	Clavicle Scapula (spine and acromion)	Humerus (lateral side about half-way down—deltoid tubercle)	Abducts upper arm Assists in flexion and extension of upper arm	Axillary nerve
Coracobrachialis	Scapula (coracoid process)	Humerus (middle third, medial surface)	Adduction; assists in flexion and medial rotation of arm	Musculocutaneous nerve
Supraspinatus†	Scapula (supraspinous fossa)	Humerus (greater tubercle)	Assists in abducting arm	Suprascapular nerve
Teres minor†	Scapula (axillary border)	Humerus (greater tubercle)	Rotates arm outward	Axillary nerve
Teres major	Scapula (lower part, axillary border)	Humerus (upper part, anterior surface)	Assists in extension, adduction, and medial rotation of arm	Lower subscapular nerve
Infraspinatus†	Scapula (infraspinatus border)	Humerus (greater tubercle)	Rotates arm outward	Suprascapular nerve
Subscapularis†	Scapula (subscapular fossa)	Humerus (lesser tubercle)	Medial rotation	Suprascapular nerve

*Axial muscles originate on the axial skeleton.
Scapular muscles originate on the scapula.
†Muscles of the rotator cuff

48

MUSCLES THAT MOVE THE FOREARM

Muscle	Origin	Insertion	Function	Nerve Supply
FLEXORS				
Biceps brachii	Scapula (supraglenoid tuberosity) Scapula (coracoid)	Radius (tubercle at proximal end)	Flexes supinated forearm Supinates forearm and hand	Musculocutaneous nerve
Brachialis	Humerus (distal half, anterior surface)	Ulna (front of coronoid process)	Flexes pronated forearm	Musculocutaneous nerve
Brachioradialis	Humerus (above lateral epicondyle)	Radius (styloid process)	Flexes semipronated or semi-supinated forearm; supinates forearm and hand	Radial nerve
EXTENSOR				
Triceps brachii	Scapula (infraglenoid tuberosity) Humerus (posterior surface—lateral head above radial groove; medial head, below)	Ulna (olecranon process)	Extends lower arm	Radial nerve
PRONATORS				
Pronator teres	Humerus (medial epicondyle) Ulna (coronoid process)	Radius (middle third of lateral surface)	Pronates and flexes forearm	Median nerve
Pronator quadratus	Ulna (distal fourth, anterior surface)	Radius (distal fourth, anterior surface)	Pronates forearm	Median nerve
SUPINATOR				
Supinator	Humerus (lateral epicondyle) Ulna (proximal fifth)	Radius (proximal third)	Supinates forearm	Radial nerve

49

MUSCLES THAT MOVE THE WRIST, HAND, AND FINGERS

Muscle	Origin	Insertion	Function	Nerve Supply
EXTRINSIC				
Flexor carpi radialis	Humerus (medial epicondyle)	Second metacarpal (base of)	Flexes hand / Flexes forearm	Median nerve
Palmaris longus	Humerus (medial epicondyle)	Fascia of palm	Flexes hand	Median nerve
Flexor carpi ulnaris	Humerus (medial epicondyle) / Ulna (proximal two thirds)	Pisiform bone / Third, fourth, and fifth metacarpals	Flexes hand / Adducts hand	Ulnar nerve
Extensor carpi radialis longus	Humerus (ridge above lateral epicondyle)	Second metacarpal (base of)	Extends hand / Abducts hand (moves toward thumb side when hand supinated)	Radial nerve
Extensor carpi radialis brevis	Humerus (lateral epicondyle)	Second, third metacarpals (bases of)	Extends hand	Radial nerve
Extensor carpi ulnaris	Humerus (lateral epicondyle) / Ulna (proximal three fourths)	Fifth metacarpal (base of)	Extends hand / Adducts hand (moves toward little finger side when hand supinated)	Radial nerve
Flexor digitorum profundus	Ulna (anterior surface)	Distal phalanges (fingers 2 to 5)	Flexes distal interphalangeal joints	Median and ulnar nerves
Flexor digitorum superficialis	Humerus (medial epicondyle) / Radius / Ulna (coronoid process)	Tendons of fingers	Flexes fingers	Median nerve

	Humerus (lateral epicondyle)	Phalanges (fingers 2 to 5)	Extends fingers	Radial nerve
Extensor digitorum	Humerus (lateral epicondyle)	Phalanges (fingers 2 to 5)	Extends fingers	Radial nerve
INTRINSIC				
Opponens pollicis	Trapezium	Thumb metacarpal	Opposes thumb to fingers	Median nerve
Abductor pollicis brevis	Trapezium	Proximal phalanx of thumb	Abducts thumb	Median nerve
Adductor pollicis	Second and third metacarpals, Trapezoid, Capitate	Proximal phalanx of thumb	Adducts thumb	Ulnar nerve
Flexor pollicis brevis	Flexor retinaculum	Proximal phalanx of thumb	Flexes thumb	Median and ulnar nerves
Abductor digiti minimi	Pisiform	Proximal phalanx of fifth finger (base of)	Abducts fifth finger / Flexes fifth finger	Ulnar nerve
Flexor digiti minimi brevis	Hamate	Proximal and middle phalanx of fifth finger	Flexes fifth finger	Ulnar nerve
Opponens digiti minimi	Hamate, Flexor retinaculum	Fifth metacarpal	Opposes fifth finger slightly	Ulnar nerve
Interosseous (palmar and dorsal)	Metacarpals	Proximal phalanges	Adducts second, fourth, fifth fingers (palmar) / Abducts second, third, fourth fingers (dorsal)	Ulnar nerve
Lumbricales	Tendons of flexor digitorum profundus	Phalanges (2 to 5)	Flexes proximal phalanges (2 to 5) / Extends middle and distal phalanges (2 to 5)	Median nerve (phalanges 2 and 3) / Ulnar nerve (phalanges 4 and 5)

MUSCLES THAT MOVE THE THIGH

Muscle	Origin	Insertion	Function	Nerve Supply
ILIOPSOAS (Iliacus, psoas major, and psoas minor)	Ilium (iliac fossa) Vertebrae (bodies of twelfth thoracic to fifth lumbar)	Femur (lesser trochanter)	Flexes thigh Flexes trunk (when femur acts as origin)	Femoral and second to fourth lumbar nerves
RECTUS FEMORIS	Ilium (anterior, inferior spine)	Tibia (by way of patellar tendon)	Flexes thigh Extends lower leg	Femoral nerve
GLUTEAL GROUP				
Maximus	Ilium (crest and posterior surface) Sacrum and coccyx (posterior surface) Sacrotuberous ligament	Femur (gluteal tuberosity) Iliotibial tract	Extends thigh—rotates outward	Inferior gluteal nerve
Medius	Ilium (lateral surface)	Femur (greater trochanter)	Abducts thigh—rotates outward; stabilizes pelvis on femur	Superior gluteal nerve
Minimus	Ilium (lateral surface)	Femur (greater trochanter)	Abducts thigh; stabilizes pelvis on femur Rotates thigh medially	Superior gluteal nerve

Muscle	Origin	Insertion	Action	Nerve
TENSOR FASCIAE LATAE	Ilium (anterior part of crest)	Tibia (by way of iliotibial tract)	Adducts thigh Tightens iliotibial tract	Superior gluteal nerve
ADDUCTOR GROUP				
Brevis	Pubic bone	Femur (linea aspera)	Adducts thigh	Obturator nerve
Longus	Pubic bone	Femur (linea aspera)	Adducts thigh	Obturator nerve
Magnus	Pubic bone	Femur (linea aspera)	Adducts thigh	Obturator nerve
GRACILIS	Pubic bone (just below symphysis)	Tibia (medial surface behind sartorius)	Adducts thigh and flexes and adducts leg	Obturator nerve

MUSCLES THAT MOVE THE LOWER LEG

Muscle	Origin	Insertion	Function	Nerve Supply
QUADRICIPS FEMORIS GROUP				
Rectus femoris	Ilium (anterior inferior spine)	Tibia (by way of patellar tendon)	Flexes thigh Extends leg	Femoral nerve
Vastus lateralis	Femur (linea aspera)	Tibia (by way of patellar tendon)	Extends leg	Femoral nerve
Vastus medialis	Femur	Tibia (by way of patellar tendon)	Extends leg	Femoral nerve
Vastus intermedius	Femur (anterior surface)	Tibia (by way of patellar tendon)	Extends leg	Femoral nerve
SARTORIUS	Coxal (anterior, superior iliac spines)	Tibia (medial surface of upper end of shaft)	Adducts and flexes leg Permits crossing of legs tailor fashion	Femoral nerve
HAMSTRING GROUP				
Biceps femoris	Ischium (tuberosity) Femur (linea aspera)	Fibula (head of)	Flexes leg	Hamstring nerve (branch of sciatic nerve) Hamstring nerve
Semitendinosus	Ischium (tuberosity)	Tibia (proximal end, medial surface)	Extends thigh	Hamstring nerve
Semimembranosus	Ischium (tuberosity)	Tibia (medial condyle)	Extends thigh	Hamstring nerve

MUSCLES THAT MOVE THE FOOT

Muscle	Origin	Insertion	Function	Nerve Supply
EXTRINSIC				
Tibialis anterior	Tibia (lateral condyle of upper body)	Tarsal (first cuneiform) Metatarsal (base of first)	Flexes foot Inverts foot	Common and deep peroneal nerves
Gastrocnemius	Femur (condyles)	Tarsal (calcaneus by way of Achilles tendon)	Extends foot Flexes lower leg	Tibial nerve (branch of sciatic nerve)
Soleus	Tibia (underneath gastrocnemius) Fibula	Tarsal (calcaneus by way of Achilles tendon)	Extends foot (plantar flexion)	Tibial nerve
Peroneus longus	Tibia (lateral condyle) Fibula (head and shaft)	First cuneiform Base of first metatarsal	Extends foot (plantar flexion) Everts foot	Common peroneal nerve
Peroneus brevis	Fibula (lower two thirds of lateral surface of shaft)	Fifth metatarsal (tubercle, dorsal surface)	Everts foot Flexes foot	Superficial peroneal nerve
Peroneus tertius	Fibula (distal third)	Fourth and fifth metatarsals (bases of)	Flexes foot Everts foot	Deep peroneal nerve
Extensor digitorum longus	Tibia (lateral condyle) Fibula (anterior surface)	Second and third phalanges (four lateral toes)	Dorsiflexion of foot; extension of toes	Deep peroneal nerve

(continued)

55

MUSCLES THAT MOVE THE FOOT—con't'd

Muscle	Origin	Insertion	Function	Nerve Supply
INTRINSIC				
Lumbricales	Tendons of flexor digitorum longus	Phalanges (2 to 5)	Flex proximal phalanges Extend middle and distal phalanges	Lateral and medial plantar nerve
Flexor digiti minimi brevis	Fifth metatarsal	Proximal phalanx of fifth toe	Flexes fifth (small) toe	Lateral plantar nerve
Flexor hallucis brevis	Cuboid Medial and lateral cuneiform	Proximal phalanx of first (great) toe	Flexes first (great) toe	Medial and lateral plantar nerve
Flexor digitorum brevis	Calcaneous Plantar fascia	Middle phalanges of toes (2 to 5)	Flexes toes two through five	Medial plantar nerve
Abductor digiti minimi	Calcaneous	Proximal phalanx of fifth (small) toe	Abducts fifth (small) toe Flexes fifth toe	Lateral plantar nerve
Abductor hallucis	Calcaneous	First (great) toe	Abducts first (great) toe	Medial plantar nerve

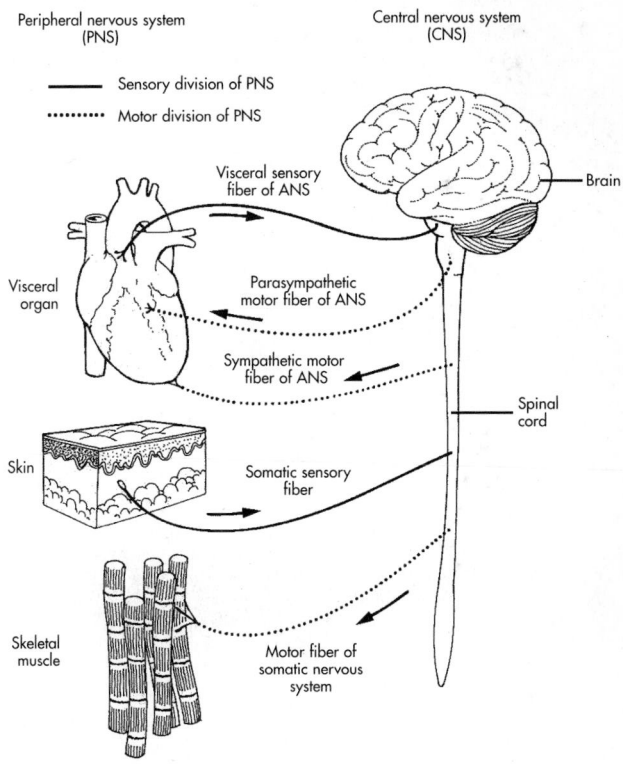

Organization of the nervous system. Visceral organs (generally within the ventral body cavity) are served by motor fibers of the autonomic nervous system (ANS) and by visceral sensory fibers. The somata (limbs and body walls) are served by motor fibers of the somatic nervous system and by somatic sensory fibers. Arrows indicate the direction of nerve impulses.

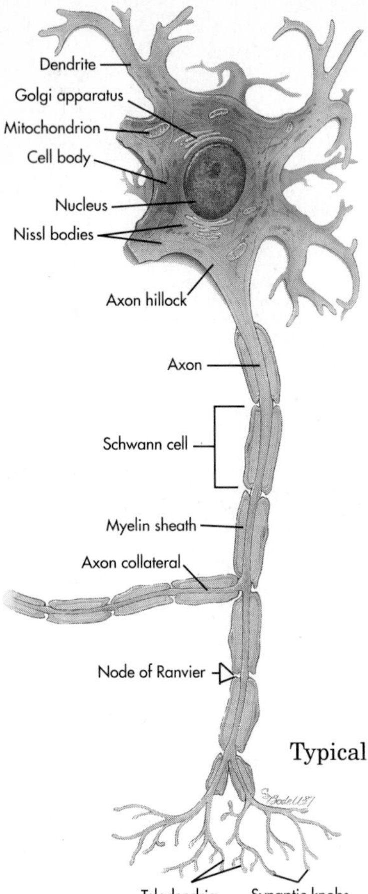

Dendrite

Golgi apparatus

Mitochondrion

Cell body

Nucleus

Nissl bodies

Axon hillock

Axon

Schwann cell

Myelin sheath

Axon collateral

Node of Ranvier

Typical multipolar neuron.

Telodendria   Synaptic knobs

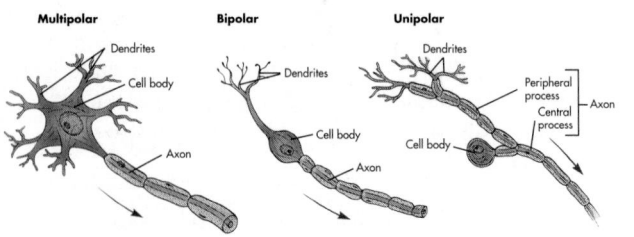

Multipolar

Dendrites

Cell body

Axon

Bipolar

Dendrites

Cell body

Axon

Unipolar

Dendrites

Peripheral process

Central process

Axon

Cell body

Structural classification of neurons:

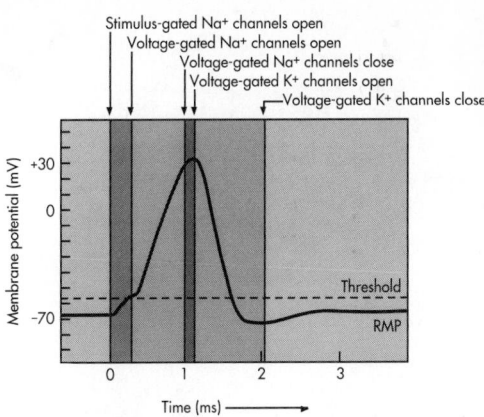

Stimulus-gated Na⁺ channels open
Voltage-gated Na⁺ channels open
Voltage-gated Na⁺ channels close
Voltage-gated K⁺ channels open
Voltage-gated K⁺ channels close

STEPS OF THE MECHANISM THAT PRODUCES AN ACTION POTENTIAL

Step	Description
1	A stimulus triggers Na⁺ channels to open and allow inward Na⁺ diffusion. This causes the membrane to depolarize.
2	As the threshold potential is reached, additional Na⁺ channels open and even more Na⁺ enters the cell—causing the membrane to depolarize further.
3	The magnitude of the action potential peaks (at +30 mV) when Na⁺ channels close.
4	Repolarization begins when K⁺ channels open, allowing outward diffusion of K⁺.
5	After a brief period of hyperpolarization, the resting potential is restored by the sodium-potassium pump and the return of ion channels to their resting state.

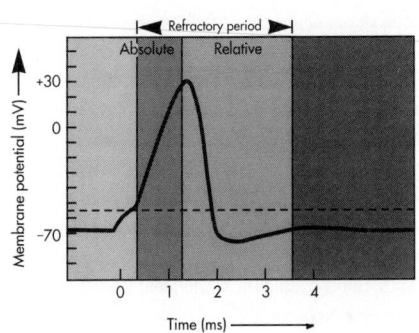

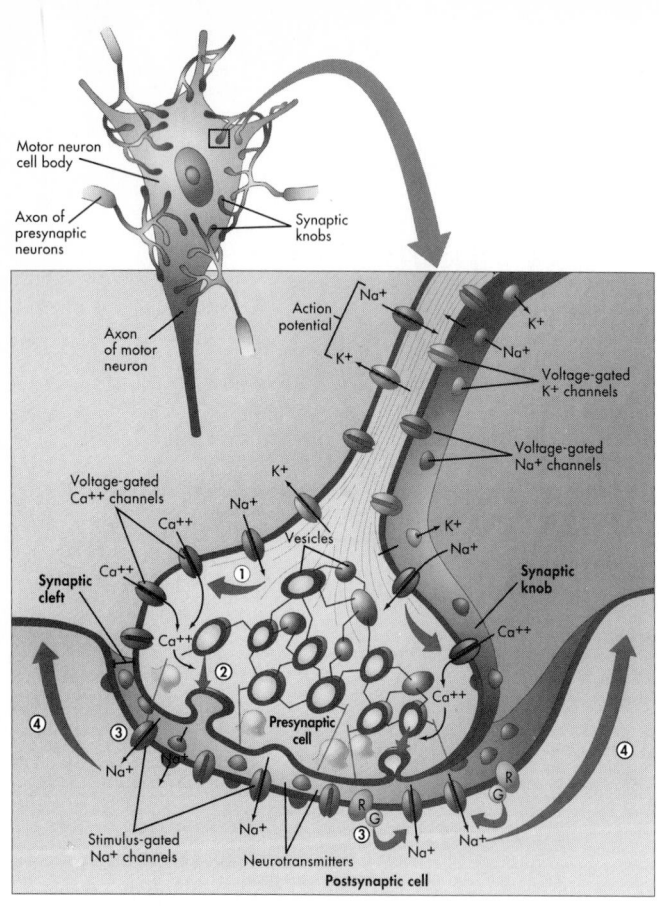

Diagram shows detail of synaptic knob, or axon terminal, of presynaptic neuron, the plasma membrane of a postsynaptic neuron, and a synaptic cleft. On the arrival of an action potential at a synaptic knob, voltage-gated calcium channels open and allow extracellular Ca^{++} to diffuse into the presynaptic cell (step1). In step 2, the Ca^{++} triggers the rapid exocytosis of neurotransmitter molecules from vesicles in the knob. In step 3, neurotransmitter diffuses into the synaptic cleft and binds to receptor molecules in the plasma membrane of the postsynaptic neuron. The postsynaptic receptors directly or indirectly trigger the opening of stimulus-gated ion channels, initiating a local potential in the postsynaptic neuron. In step 4, the local potential may move toward the axon, where an action potential may begin.

EXAMPLES OF NEUROTRANSMITTERS

Neurotransmitter	Location*	Function*
ACETYLCHOLINE	Junctions with motor effectors (muscles, glands); many parts of brain	Excitatory or inhibitory; involved in memory
AMINES		
Serotonin	Several regions of the CNS	Mostly inhibitory; involved in moods and emotions, sleep
Histamine	Brain	Mostly excitatory; involved in emotions and regulation of body temperature and water balance
Dopamine	Brain; autonomic system	Mostly inhibitory; involved in emotions/moods and in regulating motor control
Epinephrine	Several areas of the CNS and in the sympathetic division of ANS	Excitatory or inhibitory; acts as a hormone when secreted by sympathetic neurosecretory cells of the adrenal gland
Norepinephrine	Several areas of the CNS and in the sympathetic division of ANS	Excitatory or inhibitory; regulates sympathetic effectors; in brain, involved in emotional responses
AMINO ACIDS		
Glutamate (glutamic acid)	CNS	Excitatory; most common excitatory neurotransmitter in CNS
Gamma-aminobutyric acid (GABA)	Brain	Inhibitory; most common inhibitory neurotransmitter in brain
Glycine	Spinal cord	Inhibitory; most common inhibitory neurotransmitter in spinal cord

(continued)

EXAMPLES OF NEUROTRANSMITTERS—cont'd

Neurotransmitter	Location*	Function*
NEUROPEPTIDES		
Vasoactive intestinal peptide (VIP)	Brain; some ANS and sensory fibers; retina; gastrointestinal tract	Function in nervous system uncertain
Cholecystokinin (CCK)	Brain; retina	Function in nervous system uncertain
Substance P	Brain, spinal cord, sensory pain pathways; gastrointestinal tract	Mostly excitatory; transmits pain information
Enkephalins	Several regions of CNS; retina; intestinal tract	Mostly inhibitory; act like opiates to block pain
Endorphins	Several regions of CNS; retina; intestinal tract	Mostly inhibitory; act like opiates to block pain

*These are examples only; most of these neurotransmitters are also found in other locations, and many have additional functions.

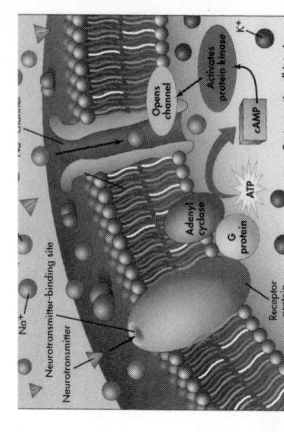

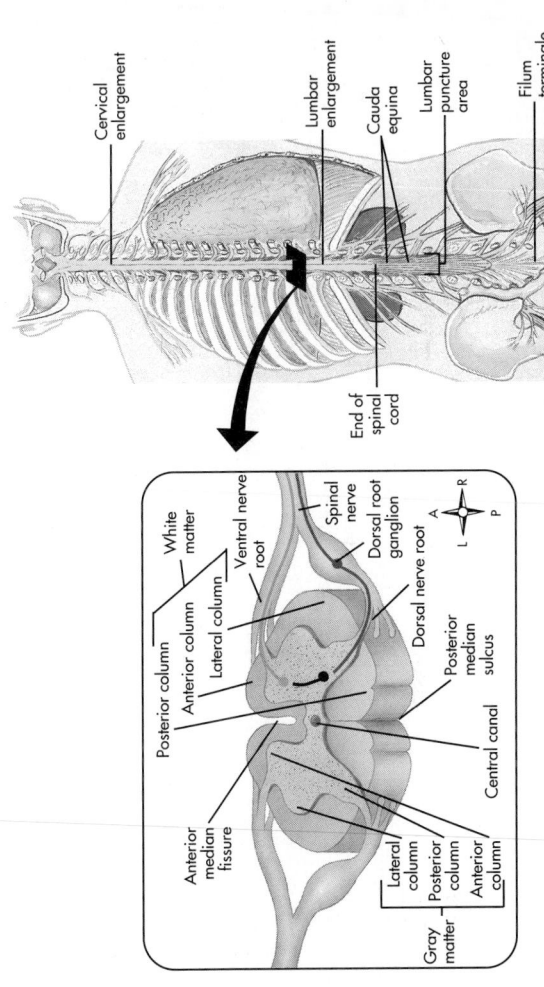

Spinal cord. The inset shows a transverse section of the spinal cord in the broader view.

MAJOR ASCENDING TRACTS OF SPINAL CORD

Name	Function	Location	Origin*	Termination†
Lateral spinothalamic	Pain, temperature, and crude touch opposite side	Lateral white columns	Posterior gray column opposite side	Thalamus
Anterior spinothalamic	Crude touch and pressure	Anterior white columns	Posterior gray column opposite side	Thalamus
Fasciculi gracilis and cuneatus	Discriminating touch and pressure sensations, including vibration, stereognosis, and two-point discrimination; also conscious kinesthesia	Posterior white columns	Spinal ganglia same side	Medulla
Anterior and posterior spinocerebellar	Unconscious kinesthesia	Lateral white columns	Anterior or posterior gray column	Cerebellum

*Location of cell bodies of neurons from which axons of tract arise.
†Structure in which axons of tract terminate.

MAJOR DESCENDING TRACTS OF SPINAL CORD

Name	Function	Location	Origin*	Termination†
Lateral corticospinal (or crossed pyramidal)	Voluntary movement, contraction of individual or small groups of muscles, particularly those moving hands, fingers, feet, and toes of opposite side	Lateral white columns	Motor areas or cerebral cortex opposite side from tract location in cord	Lateral or anterior gray columns
Anterior corticospinal (direct pyramidal)	Same as lateral corticospinal except mainly muscles of same side	Anterior white columns	Motor cortex but on same side as location in cord	Lateral or anterior gray columns
Lateral reticulospinal	Mainly facilitatory influence on motor neurons to skeletal muscles	Lateral white columns	Reticular formation, midbrain pons, and medulla	Lateral or anterior gray columns
Medial reticulospinal	Mainly inhibitory influence on motor neurons to skeletal muscles	Anterior white columns	Reticular formation, medulla mainly	Lateral or anterior gray columns
Rubrospinal	Coordination of body movement and posture	Lateral white columns	Red nucleus (of midbrain)	Lateral or anterior gray columns

*Location of cell bodies of neurons from which axons of tract arise.
†Structure in which axons of tract terminate.

65

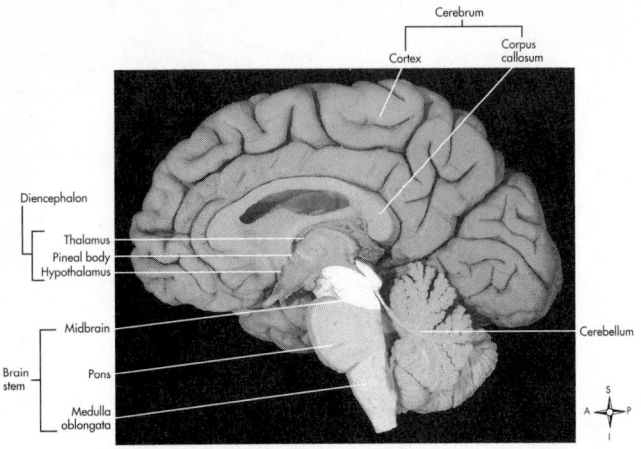

Divisions of the brain as seen in a midsagittal section.

FUNCTIONS OF MAJOR DIVISIONS OF THE BRAIN

Brain Area	Function
BRAIN STEM	
Medulla oblongata	Two-way conduction pathway between the spinal cord and higher brain centers; cardiac, respiratory, and vasomotor control center
Pons	Two-way conduction pathway between areas of the brain and other regions of the body; influences respiration
Midbrain	Two-way conduction pathway; relay for visual and auditory impulses
DIENCEPHALON	
Hypothalamus	Regulation of body temperature, water balance, sleep-cycle control, appetite, and sexual arousal
Thalamus	Sensory relay station from various body areas to cerebral cortex; emotions and alerting or arousal mechanisms
CEREBELLUM	Muscle coordination; maintenance of equilibrium and posture
CEREBRUM	Sensory perception, emotions, willed movements, consciousness, and memory

SPINAL NERVES AND PERIPHERAL BRANCHES

Spinal Nerves	Plexuses Formed From Anterior Rami	Spinal Nerve Branches From Plexuses	Parts Supplied
CERVICAL			
1		Lesser occipital	Sensory to back of head, front of neck, and upper part of shoulder; motor to numerous neck muscles
2		Greater auricular	
3	Cervical plexus	Cutaneous nerve of neck	
4		Supraclavicular nerves	
		Branches to muscles	
CERVICAL		Phrenic nerve	Diaphragm
5		Suprascapular and dorsoscapular	Superficial muscles* of scapula
6		Thoracic nerves, medial and lateral branches	Pectoralis major and minor
7	Brachial plexus	Long thoracic nerve	Serratus anterior
8		Thoracodorsal	Latissimus dorsi
		Subscapular	Subscapular and teres major muscles
		Axillary (circumflex)	Deltoid and teres minor muscles and skin over deltoid
		Musculocutaneous	Muscles of front of arm (biceps brachii, coracobrachialis, and brachialis) and skin on outer side of forearm
THORACIC (OR DORSAL)			
1		Ulnar	Flexor carpi ulnaris and part of flexor digitorum profundus; some of muscles of hand; sensory to medial side of hand, little finger, and medial half of fourth finger
2			
3			
4		Median	Rest of muscles of front of forearm and hand; sensory to skin of palmar surface of thumb, index, and middle fingers
5	No plexus formed; branches run directly to intercostal muscles and skin of thorax		
6			
7		Radial	Triceps muscle and muscles of back of forearm; sensory to skin of back of forearm and hand
8			
9			
10		Medial cutaneous	Sensory to inner surface of arm and forearm
11			
12			

(continued)

67

Spinal Nerves	Plexuses Formed From Anterior Rami	Spinal Nerve Branches From Plexuses		Parts Supplied
LUMBAR				
1		Iliohypogastric	Sometimes fused	Sensory to anterior abdominal wall
		Ilioinguinal		Sensory to anterior abdominal wall and external genitalia; motor to muscles of abdominal wall
2		Genitofemoral		Sensory to skin of external genitalia and inguinal region
3		Lateral femoral cutaneous		Sensory to outer side of thigh
4		Femoral		Motor to quadriceps, sartorius, and iliacus muscles; sensory to front of thigh and medial side of lower leg (saphenous nerve)
5	Lumbosacral plexus			Motor to adductor muscles of thigh
SACRAL				
1		Obturator		Motor to muscles of calf of leg; sensory to skin of calf of leg and sole of foot
2		Tibial† (medial popliteal)		Motor to evertors and dorsiflexors of foot; sensory to lateral surface of leg and dorsal surface of foot
3		Common peroneal (lateral popliteal)		Motor to muscles of back of thigh
4		Nerves to hamstring muscles		Motor to buttock muscles and tensor fasciae latae
5		Gluteal nerves		Sensory to skin of buttocks, posterior surface of thigh, and leg
COCCYGEAL	Coccygeal plexus	Posterior femoral cutaneous		
1		Pudendal nerve		Motor to perineal muscles; sensory to skin of perineum

*Although nerves to muscles are considered motor, they do contain some sensory fibers that transmit proprioceptive impulses.

†Sensory fibers from the tibial and peroneal nerves unite to form the *medial cutaneous* (or *sural*) *nerve* that supplies the calf of the leg and the lateral surface of the foot. In the thigh the tibial and common peroneal nerves are usually enclosed in a single sheath to form the *sciatic nerve,* the largest nerve in the body with a width of approximately 3/4 of an inch. About two thirds of the way down the posterior part of the thigh, it divides into its component parts. Branches of the sciatic nerve extend into the hamstring muscles.

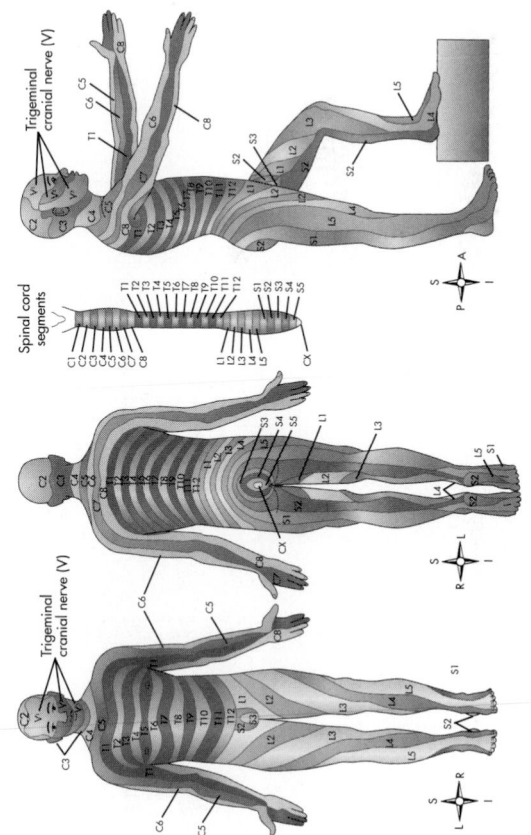

Dermatome distribution of spinal nerves.

| Nerve | Sensory Fibers | | | Motor Fibers | | Functions |
	Receptors	Cell Bodies	Termination	Cell Bodies	Termination	
I OLFACTORY	Nasal mucosa	Nasal mucosa	Olfactory bulbs (new relay of neurons to olfactory cortex)			Sense of smell
II OPTIC	Retina	Retina	Nucleus in thalamus (lateral geniculate); some fibers terminate in superior colliculus of midbrain			Vision
III OCULOMOTOR	External eye muscles except superior oblique and lateral rectus	Trigeminal ganglion	Midbrain (oculomotor nucleus)	Midbrain (oculomotor nucleus)	External eye muscles except superior oblique and lateral rectus; autonomic fibers terminate in ciliary ganglion and then to ciliary and iris muscles	Eye movements, regulation of size of pupil, accommodation (for near vision), proprioception (muscle sense)
IV TROCHLEAR	Superior oblique (proprioceptive)	Trigeminal ganglion	Midbrain	Midbrain	Superior oblique muscle of eye	Eye movements, proprioception
V TRIGEMINAL	Skin and mucosa of head, teeth	Trigeminal ganglion	Pons (sensory nucleus)	Pons (motor nucleus)	Muscles of mastication	Sensations of head and face, chewing movements, proprioception
VI ABDUCENS	Lateral rectus (proprioceptive)	Trigeminal ganglion	Pons	Pons	Lateral rectus muscle of eye	Abduction of eye, proprioception

(continued)

Nerve	Sensory Fibers			Motor Fibers		Functions
	Receptors	Cell Bodies	Termination	Cell Bodies	Termination	
VII FACIAL	Taste buds of anterior two thirds of tongue	Geniculate ganglion	Medulla (nucleus solitarius)	Pons	Superficial muscles of face and scalp; autonomic fibers to salivary and lacrimal glands	Facial expressions, secretion of saliva and tears, taste
VIII VESTIBULOCOCHLEAR						
Vestibular Branch	Semicircular canals and vestibule (utricle and saccule)	Vestibular ganglion	Pons and medulla (vestibular nuclei)			Balance or equilibrium sense
Cochlear or Auditory Branch	Organ of Corti in cochlear duct	Spiral ganglion	Pons and medulla (cochlear nuclei)			Hearing
IX GLOSSO-PHARYNGEAL	Pharynx; taste buds and other receptors of posterior one third of tongue	Jugular and petrous ganglia	Medulla (nucleus solitarius)	Medulla (nucleus ambiguus)	Muscles of pharynx	Sensations of tongue, swallowing movements, secretion of saliva, aid in reflex control of blood pressure and respiration
	Carotid sinus and carotid body	Jugular and petrous ganglia	Medulla (respiratory and vasomotor centers)	Medulla at junction of pons (nucleus salivatorius)	Otic ganglion and then to parotid salivary gland	
X VAGUS	Pharynx, larynx, carotid body, and thoracic and abdominal viscera	Jugular and nodose ganglia	Medulla (nucleus solitarius), pons (nucleus of fifth cranial nerve)	Medulla (dorsal motor nucleus)	Ganglia of vagal plexus and then to muscles of pharynx, larynx, and autonomic fibers to thoracic and abdominal viscera	Sensations and movements of organs supplied; for example, slows heart, increases peristalsis, and contracts muscles for voice production

(continued)

	Sensory Fibers			Motor Fibers		
Nerve	Receptors	Cell Bodies	Termination	Cell Bodies	Termination	Functions
XI ACCESSORY	Trapezius and sternocleidomastoid (proprioceptive)	Upper, cervical ganglia	Spinal cord	Medulla (dorsal motor nucleus of vagus and nucleus ambiguus)	Muscles of thoracic and abdominal viscera (autonomic) and pharynx and larynx	Shoulder movements, turning movements of head, movements of viscera, voice production, proprioception
				Anterior gray column of first five or six cervical segments of spinal cord	Trapezius and sternocleidomastoid muscle	
XII HYPO-GLOSSAL	Tongue muscles (proprioceptive)	Trigeminal ganglion	Medulla (hypoglossal nucleus)	Medulla (hypoglossal nucleus)	Muscles of tongue and throat	Tongue movements, proprioception

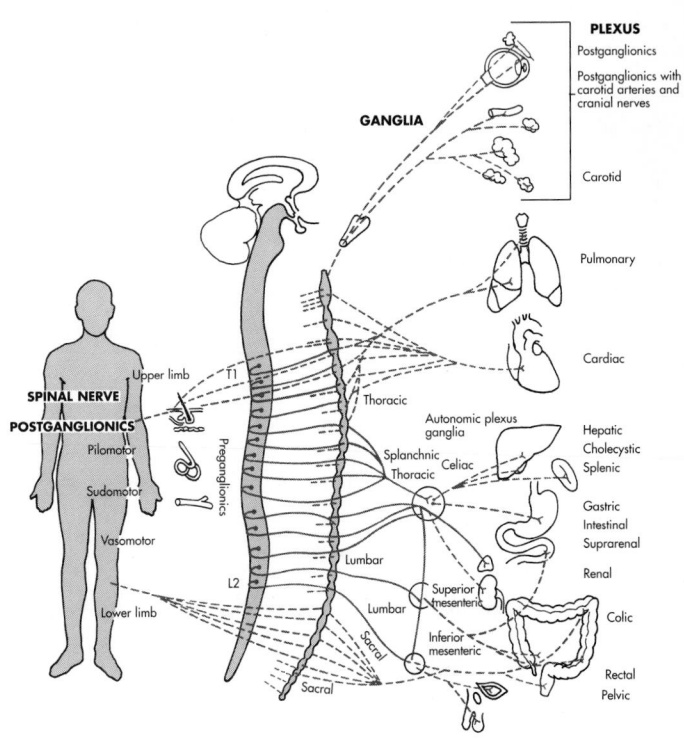

Schematic representation of the sympathetic nervous system.

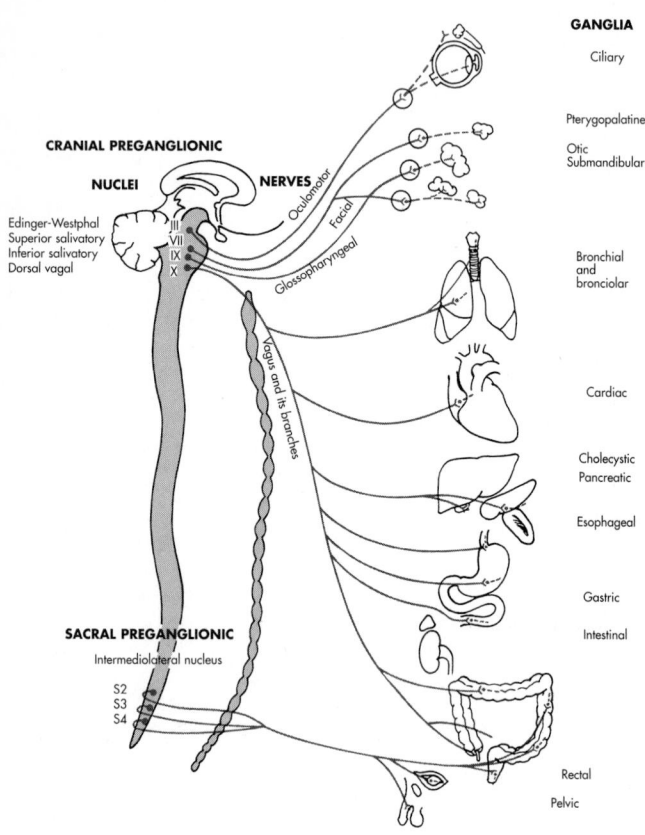

Schematic representation of the parasympathetic nervous system.

COMPARISON OF STRUCTURAL FEATURES OF THE SYMPATHETIC AND PARASYMPATHETIC PATHWAYS

Neurons	Sympathetic	Parasympathetic
PREGANGLIONIC NEURONS		
Dendrites and cell bodies	In lateral gray columns of thoracic and first four lumbar segments of spinal cord	In nuclei of brain stem and in lateral gray columns of sacral segments of cord
Axons	In anterior roots of spinal nerves to spinal nerves (thoracic and first four lumbar), to and through white rami to terminate in sympathetic ganglia at various levels or to extend through sympathetic ganglia, to and through splanchnic nerves to terminate in collateral ganglia	From brain stem nuclei through cranial nerve III to ciliary ganglion From nuclei in pons through cranial nerve VII to sphenopalatine or submaxillary ganglion From nuclei in medulla through cranial nerve IX to otic ganglion or through cranial nerves X and XI to cardiac and celiac ganglia, respectively
Distribution	Short fibers from CNS to ganglion	Long fibers from CNS to ganglion
Neurotransmitter	Acetylcholine	Acetylcholine
GANGLIA	Sympathetic chain ganglia (22 pairs); collateral ganglia (celiac, superior, and inferior mesenteric)	Terminal ganglia (in or near effector)

(continued)

75

COMPARISON OF STRUCTURAL FEATURES OF THE SYMPATHETIC AND PARASYMPATHETIC PATHWAYS—cont'd

Neurons	Sympathetic	Parasympathetic
POSTGANGLIONIC NEURONS		
Dendrites and cell bodies	In sympathetic and collateral ganglia	In parasympathetic ganglia (for example, ciliary, sphenopalatine, submaxillary, otic, cardiac, celiac) located in or near visceral effector organs
Receptors	Cholinergic (nicotinic)	Cholinergic (nicotinic)
Axons	In autonomic nerves and plexuses that innervate thoracic and abdominal viscera and blood vessels in these cavities	In short nerves to various visceral effector organs
	In gray rami to spinal nerves, to smooth muscle of skin blood vessels and hair follicles, and to sweat glands	
Distribution	Long fibers from ganglion to widespread effectors	Short fibers from ganglion to single effector
Neurotransmitter	Norepinephrine (many); acetylcholine (few)	Acetylcholine

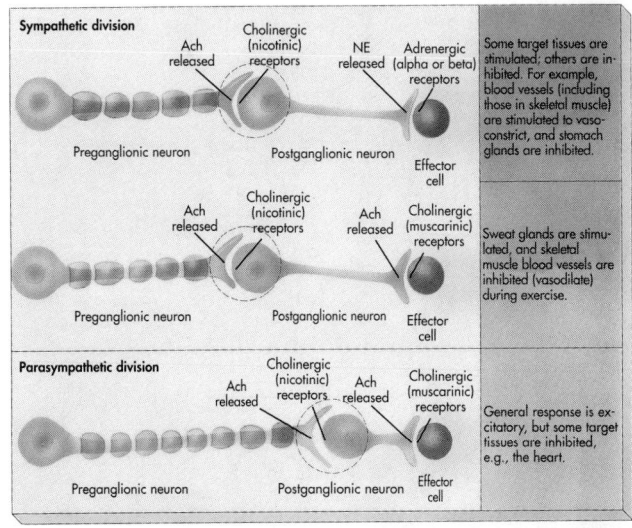

Locations of neurotransmitters and receptors of the autonomic nervous system.

AUTONOMIC FUNCTIONS

Autonomic Effector	Effect of Sympathetic Stimulation (Neurotransmitter: Norepinephrine Unless Otherwise Stated)	Effect of Parasympathetic Stimulation (Neurotransmitter: Acetylcholine)
CARDIAC MUSCLE	Increased rate and strength of contraction (beta receptors)	Decreased rate and strength of contraction
SMOOTH MUSCLE OF BLOOD VESSELS		
Skin blood vessles	Constriction (alpha receptors)	No effect
Skeletal muscle blood vessels	Dilation (beta receptors)	No effect
Coronary blood vessels	Constriction (alpha receptors) Dilation (beta receptors)	Dilation
Abdominal blood vessels	Constriction (alpha receptors)	No effect
Blood vessels of external genitals	Constriction (alpha receptors)	Dilation of blood vessels causing erection
SMOOTH MUSCLE OF HOLLOW ORGANS AND SPHINCTERS		
Bronchioles	Dilation (beta receptors)	Constriction
Digestive tract, except sphincters	Decreased peristalsis (beta receptors)	Increased peristalsis
Sphincters of digestive tract	Constriction (alpha receptors)	Relaxation
Urinary bladder	Relaxation (beta receptors)	Contraction

Urinary sphincters	Constriction (alpha receptors)	Relaxation
Reproductive ducts	Contraction (alpha receptors)	Relaxation
Eye		
Iris	Contraction of radial muscle; dilated pupil	Contraction of circular muscle; constricted pupil
Ciliary	Relaxation; accommodates for far vision	Contraction; accommodates for near vision
Hairs (pilomotor muscles)	Contraction produces goose pimples, or piloerection (alpha receptors)	No effect
GLANDS		
Sweat	Increased sweat (neurotransmitter, acetylcholine)	No effect
Lacrimal	No effect	Increased secretion of tears
Digestive (salivary, gastric, etc.)	Decreased secretion of saliva; not known for others	Increased secretion of saliva
Pancreas, including islets	Decreased secretion	Increased secretion of pancreatic juice and insulin
Liver	Increased glycogenolysis (beta receptors); increased blood sugar level	No effect
Adrenal medulla*	Increased epinephrine secretion	No effect

*Sympathetic preganglionic axons terminate in contact with secreting cells of the adrenal medulla. Thus the adrenal medulla functions, to quote someone's descriptive phrase, as a "giant sympathetic post-ganglionic neuron."

SUMMARY OF THE SYMPATHETIC "FIGHT-OR-FLIGHT" REACTION

Response	Role in Promoting Energy Use By Skeletal Muscles
Increased heart rate	Increased rate of blood flow, thus increased delivery of oxygen and glucose to skeletal muscles
Increased strength of cardiac muscle contraction	Increased rate of blood flow, thus increased delivery of oxygen and glucose to skeletal muscles
Dilation of coronary vessels of the heart	Increased delivery of oxygen and nutrients to cardiac muscle to sustain increased rate and strength of heart contractions
Dilation of blood vessels in skeletal muscles	Increased delivery of oxygen and nutrients to skeletal muscles
Constriction of blood vessels in digestive and other organs	Shunting of blood to skeletal muscles to increase oxygen and glucose delivery
Contraction of spleen and other blood reservoirs	More blood discharged into general circulation, causing increased delivery of oxygen and glucose to skeletal muscles
Dilation of respiratory airways	Increased loading of oxygen into blood
Increased rate and depth of breathing	Increased loading of oxygen into blood
Increased sweating	Increased dissipation of heat generated by skeletal muscle activity
Increased conversion of glycogen into glucose	Increased amount of glucose available to skeletal muscles

SOMATIC SENSORY RECEPTORS

Classification by Structure	By Location and Type	By Activation Stimulus	By Sensation or Function
FREE NERVE ENDINGS			
Nociceptors	Both exteroceptors and Visceroceptors—most body tissues	Almost any noxious stimulus; temperature change; mechanical	Pain; temperature; itch; tickle; stretching
Merkel discs	Exteroceptors	Light pressure; mechanical	Discriminative touch
Root hair plexuses	Exteroceptors	Hair movement; mechanical	Sense of "deflection" type movement of hair

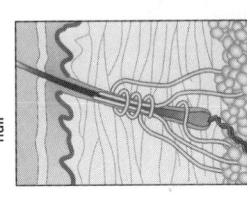

Root hair plexuses

Merkel discs

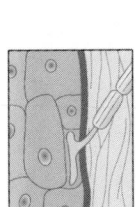

Nociceptors

(continued)

Classification by Structure	By Location and Type	By Activation Stimulus	By Sensation or Function
ENCAPSULATED NERVE ENDINGS			
Touch and pressure receptors			
Meissner's corpuscle	Exteroceptors; epidermis; hairless skin	Light pressure, mechanical	Discriminative touch; low frequency vibration
Krause's corpuscle	Mucous membranes	Mechanical; thermal?	Touch; low frequency vibration; cold?
Ruffini's corpuscle	Dermis of skin exteroceptors	Mechanical; thermal?	Crude and persistent touch; heat?
Pacinian corpuscle	Dermis of skin joint capsules	Deep Pressure, mechanical	Deep pressure; high frequency vibration; stretch
Stretch receptors			
Muscle spindles	Skeletal muscle	Stretch, mechanical	Sense of muscle length
Golgi tendon receptors	Musculotendinous junction	Force of contraction and tendon stretch, mechanical	Sense of muscle tension

Ruffini's corpuscle

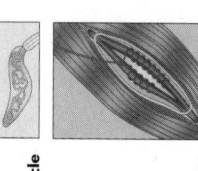

Muscle spindles

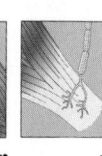

Golgi tendon receptors

Meissner's corpuscle

Krause's end bulb

Pacinian corpuscle

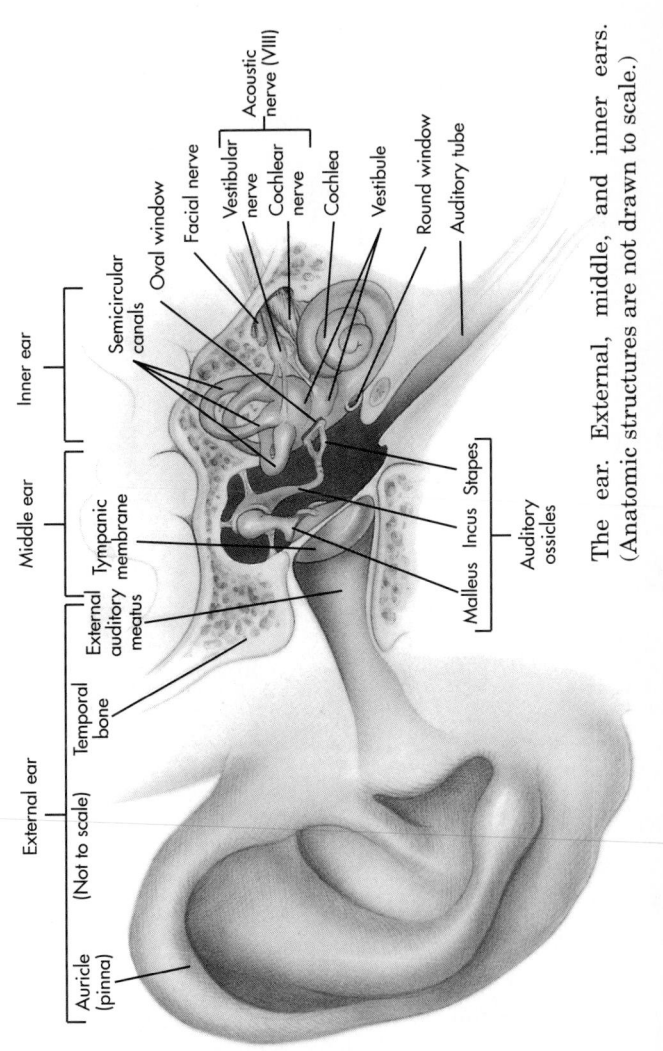

The ear. External, middle, and inner ears. (Anatomic structures are not drawn to scale.)

83

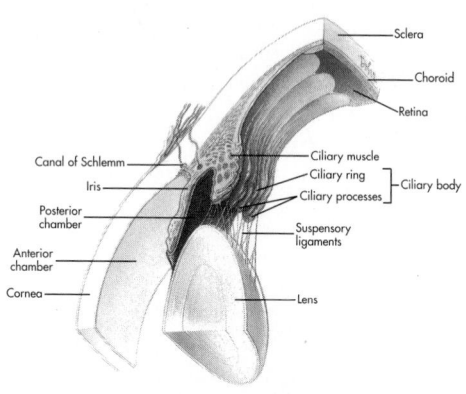

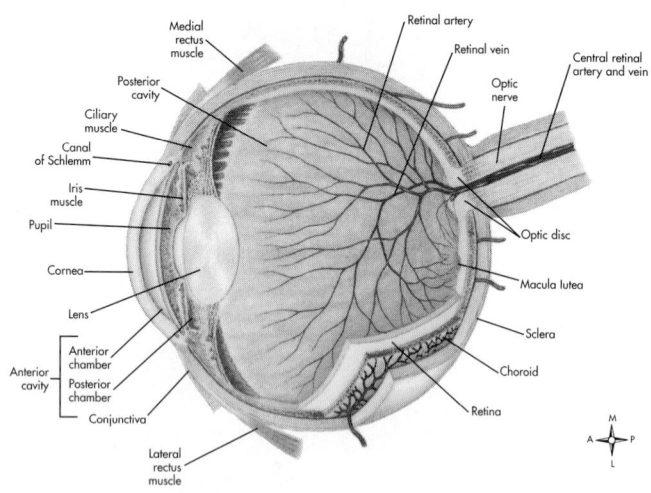

Horizontal section through the left eyeball. The eye is viewed from above. (Top) Lens, cornea, iris, and ciliary body.

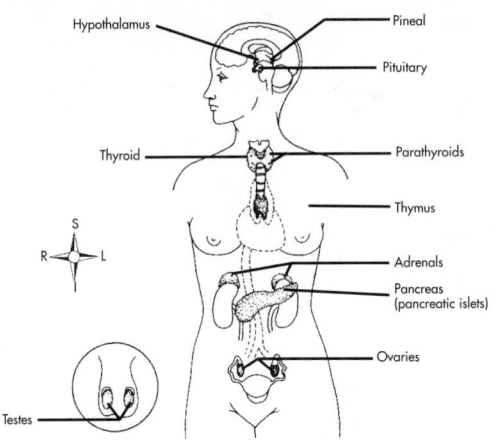

Locations of major endocrine glands.

HORMONES

STEROID
Cortisol (hydrocortisone)
Aldosterone
Estrogen
Progesterone
Testosterone

NONSTEROID
Proteins
Growth hormone (GH)
Prolactin (PRL)
Parathyroid hormone (PTH)
Calcitonin
Adrenocorticotropic hormone (ACTH)
Insulin
Glucagon
Glycoproteins
Follicle-stimulating hormone (FSH)
Luteinizing hormone (LH)
Thyroid-stimulating hormone (TSH)
Chorionic gonadotropin (CG)

NONSTEROID—cont'd
Peptides
Antidiuretic hormone (ADH)
Oxytocin
Melanocyte-stimulating hormone (MSH)
Somatostatin
Thyrotropin-releasing hormone (TRH)
Gonadotropin-releasing hormone (GnRH)
Amino acid derivatives
Amines
　Norepinephrine
　Epinephrine
　Melatonin
Iodinated amino acids
　Thyroxine (T4)
　Triiodothyronine (T3)

Chemical classification of hormones.

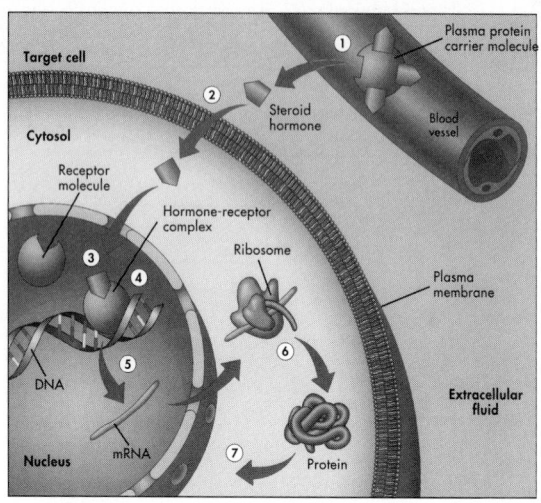

Steroid hormone mechanism.

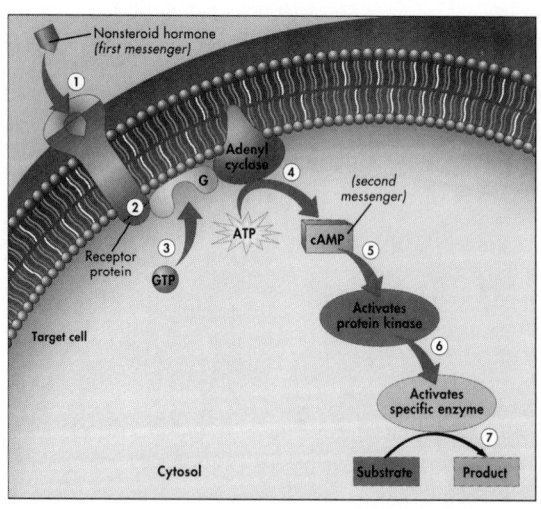

Example of a second-messenger mechanism.

HORMONES AND THEIR PRINCIPAL ACTIONS

Source	Hormone	Principal Action
Hypothalamus	Growth hormone-releasing hormone (GRH)	Stimulates secretion (release) of growth hormone
	Growth hormone-inhibiting hormone (GIH), or somatostatin	Inhibits secretion of growth hormone
	Corticotropin-releasing hormone (CRH)	Stimulates release of adrenocorticotropic hormone (ACTH)
	Thyrotropin-releasing hormone (TRH)	Stimulates release of thyroid-stimulating hormone (TSH)
	Gonadotropin-releasing hormone (GnRH)	Stimulates release of gonadotropins (FSH and LH)
	Prolactin-releasing hormone (PRH)	Stimulates secretion of prolactin
	Prolactin-inhibiting hormone (PIH)	Inhibits secretion of prolactin
Anterior pituitary gland (adeno-hypophysis)	Growth hormone (GH) (somatotropin [STH])	Promotes growth by stimulating protein anabolism and fat mobilization
	Prolactin (PRL) (lactogenic hormone)	Promotes milk secretion
	Thyroid-stimulating hormone (TSH)	Stimulates development and secretion in the thyroid gland
	Adrenocorticotropic hormone (ACTH)	Promotes development and secretion in the adrenal cortex
	Follicle-stimulating hormone (FSH)	Female: promotes development of ovarian follicle; stimulates estrogen secretion
		Male: promotes development of testis; stimulates sperm production
	Luteinizing hormone (LH)	Female: triggers ovulation; promotes development of corpus luteum
		Male: stimulates production of testosterone
	Melanocyte-stimulating hormone (MSH)	Exact function uncertain; may stimulate production of melanin pigment in skin; may maintain adrenal sensitivity
Posterior pituitary gland (neuro-hypophysis)	Antidiuretic hormone (ADH)	Promotes water retention by kidney tubules
	Oxytocin (OT)	Stimulates uterine contractions; stimulates ejection of milk into mammary ducts

(continued)

Source	Hormone	Principal Action
Thyroid gland	Triiodothyronine (T_3)	Increases rate of metabolism
	Tetraiodothyronine (T_4), or thyroxine	Increases rate of metabolism (usually converted to T_3 first)
	Calcitonin (CT)	Increases calcium storage in bone, lowering blood Ca^{++} levels
Parathyroid gland	Parathyroid hormone (PTH), or ?	Increases calcium removal from storage in bone and increases absorption of calcium by intestines, increasing blood Ca^{++} levels
Adrenal cortex	Aldosterone	Stimulates kidney tubules to conserve sodium, which, in turn, triggers the release of ADH and the resulting conservation of water by the kidney
	Cortisol (hydrocortisone)	Influences metabolism of food molecules; in large amounts, it has an anti-inflammatory effect
	Adrenal androgens	Exact role uncertain, but may support sexual function
	Adrenal estrogens	Thought to be physiologically insignificant
Adrenal medulla	Epinephrine (adrenaline)	Enhances and prolongs the effects of the sympathetic division of the ANS
	Norepinephrine	Enhances and prolongs the effects of the sympathetic division of the ANS
Pancreatic islets	Glucagon	Promotes movement of glucose from storage and into the blood
	Insulin	Promotes movement of glucose out of the blood and into cells
	Somatostatin	Can have general effects in the body, but primary role seems to be regulation of secretion of other pancreatic hormones
	Pancreatic polypeptide	Exact function uncertain, but seems to influence absorption in the digestive tract

CLASSES OF BLOOD CELLS

Cell Type	Description	Function	Life Span
Erythrocyte	7 μm in diameter; concave disk shape; entire cell stains pale pink; no nucleus	Transportation of respiratory gases (O_2 and CO_2)	105 to 120 days
Neutrophil	12-15 μm in diameter; spherical shape; multilobed nucleus; small, pink-purple staining cytoplasmic granules	Cellular defense—phagocytosis of small pathogenic microorganisms	Hours to 3 days
Basophil	11-14 μm in diameter; spherical shape; generally two-lobed nucleus; large purple staining cytoplasmic granules	Secretes heparin (anticoagulant) and histamine (important in inflammatory response)	Hours to 3 days

(continued)

CLASSES OF BLOOD CELLS—cont'd

Cell Type		Description	Function	Life Span
Eosinophil		10-12 μm in diameter; spherical shape; generally two-lobed nucleus; large orange-red staining cytoplasmic granules	Cellular defense—phagocytosis of large pathogenic microorganisms such as protozoa and parasitic worms; releases antiinflamatory substances in allergic reactions	10 to 12 days
Lymphocyte		6-9 μm in diameter; spherical shape; round (single lobe) nucleus; small lymphocytes have scant cytoplasm	Humoral defense—secretes antibodies; involved in immune system response and regulation	Days to years
Monocyte		12-17 μm in diameter; spherical shape; nucleus generally kidney-bean or "horseshoe" shaped with convoluted surface; ample cytoplasm often "steel blue" in color	Capable of migrating out of the blood to enter tissue spaces as a *macrophage*—an aggressive phagocytic cell capable of ingesting bacteria, cellular debris, and cancerous cells	Months
Platelet		2-5 μm in diameter; irregularly shaped fragments; cytoplasm contains very small pink staining granules	Releases clot activating substances and helps in formation of actual blood clot by forming platelet "plugs"	7 to 10 days

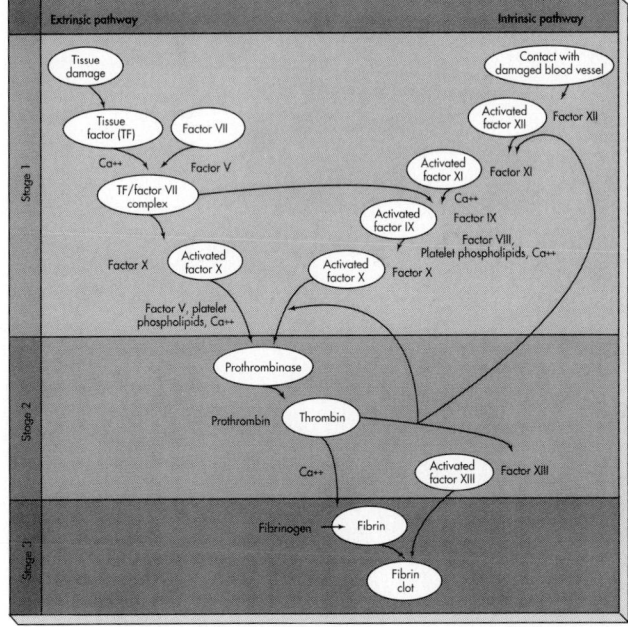

Clot formation. A, Extrinsic clotting pathway. Stage 1:
damaged tissue releases tissue factor, which with factor
VII and calcium ions activates factor X. Activated factor X,
factor V, phospholipids, and calcium ions form prothrom-
binase. Stage 2: Prothrombin is converted to thrombin by
prothrombinases. Stage 3: Fibrinogen is converted to fibrin by thrombin. Fibrin forms a clot. B, Intrinsic clotting
pathway. Stage 1: Damaged vessels cause activation of
factor XII. Activated factor XII activates factor XI, which
activates factor IX. Factor IX, with factor VIII and platelet
phospholipids, activates factor X. Activated factor X, factor V, phospholipids, and calcium ions form prothrombi-
nase. Stages 2 and 3 take the same course as the extrinsic clotting pathway.

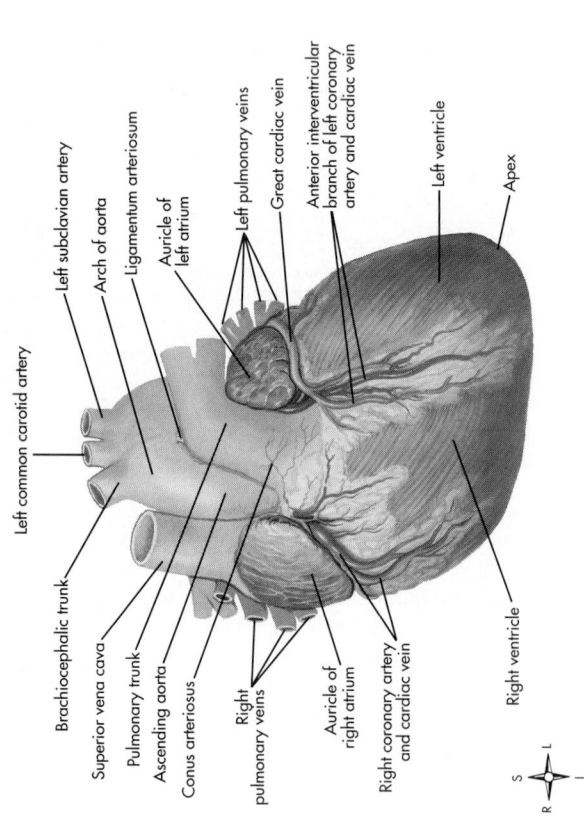

Left common carotid artery

Left subclavian artery

Arch of aorta

Ligamentum arteriosum

Auricle of left atrium

Left pulmonary veins

Great cardiac vein

Anterior interventricular branch of left coronary artery and cardiac vein

Left ventricle

Apex

Brachiocephalic trunk

Superior vena cava

Pulmonary trunk

Ascending aorta

Conus arteriosus

Right pulmonary veins

Auricle of right atrium

Right coronary artery and cardiac vein

Right ventricle

S
R — L
I

The heart and great vessels (anterior view).

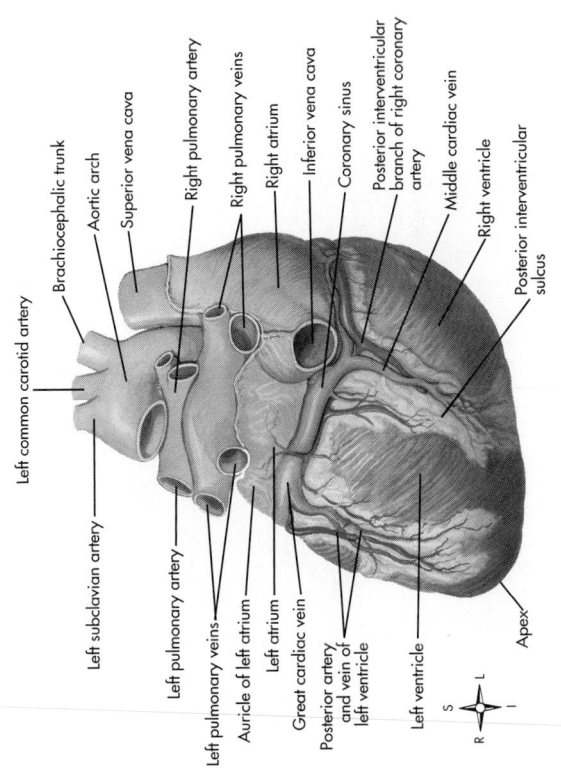

Left common carotid artery

Brachiocephalic trunk

Aortic arch

Superior vena cava

Right pulmonary artery

Right pulmonary veins

Right atrium

Inferior vena cava

Coronary sinus

Posterior interventricular branch of right coronary artery

Middle cardiac vein

Right ventricle

Posterior interventricular sulcus

Left subclavian artery

Left pulmonary artery

Left pulmonary veins

Auricle of left atrium

Left atrium

Great cardiac vein

Posterior artery and vein of left ventricle

Left ventricle

Apex

The heart and great vessels (posterior view).

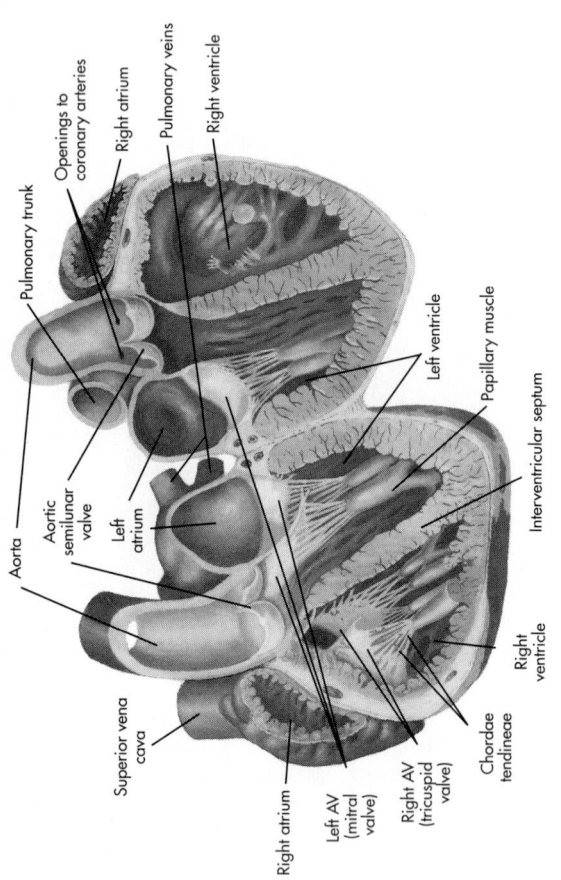

Pulmonary trunk

Openings to
coronary arteries

Right atrium

Pulmonary veins

Right ventricle

Aorta

Aortic
semilunar
valve

Left
atrium

Left ventricle

Papillary muscle

Interventricular septum

Superior vena
cava

Right
ventricle

Right atrium

Left AV
(mitral
valve)

Right AV
(tricuspid
valve)

Chordae
tendineae

Interior of the heart. The four chambers of the heart—
two atria and two ventricles—are easily seen in this figure.

94

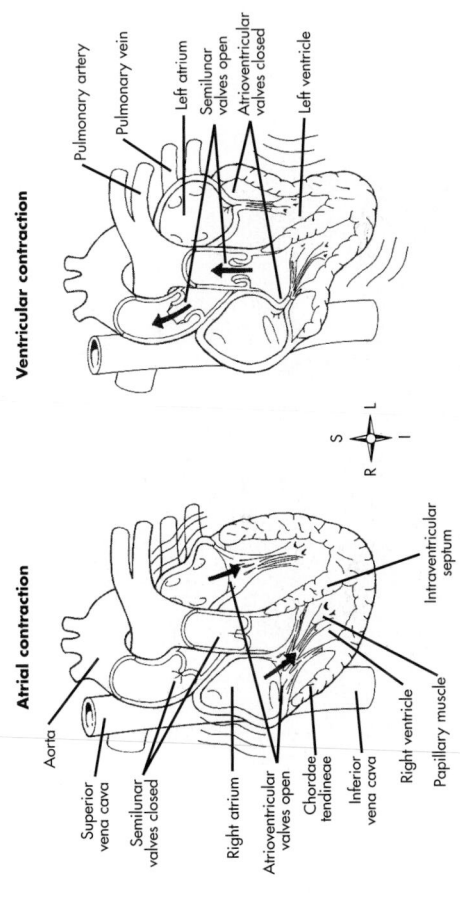

Atrial contraction

Aorta

Superior vena cava

Semilunar valves closed

Right atrium

Atrioventricular valves open

Chordae tendineae

Inferior vena cava

Right ventricle

Papillary muscle

Intraventricular septum

Ventricular contraction

Pulmonary artery

Pulmonary vein

Left atrium

Semilunar valves open

Atrioventricular valves closed

Left ventricle

S
R — L
I

Chambers and valves of the heart.

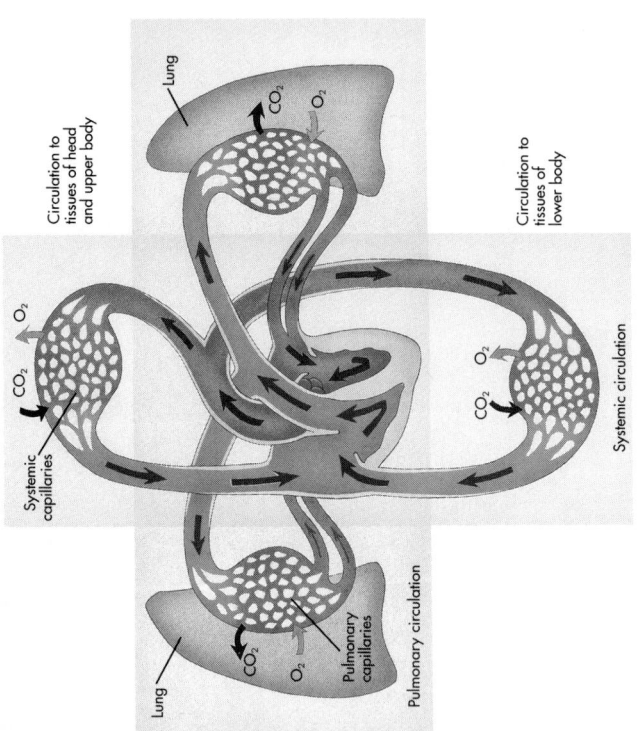

Circulation to tissues of head and upper body

Lung

CO_2

O_2

CO_2

O_2

Systemic capillaries

Circulation to tissues of lower body

Systemic circulation

CO_2

O_2

Lung

CO_2

O_2

Pulmonary capillaries

Pulmonary circulation

Blood flow through the circulatory system.

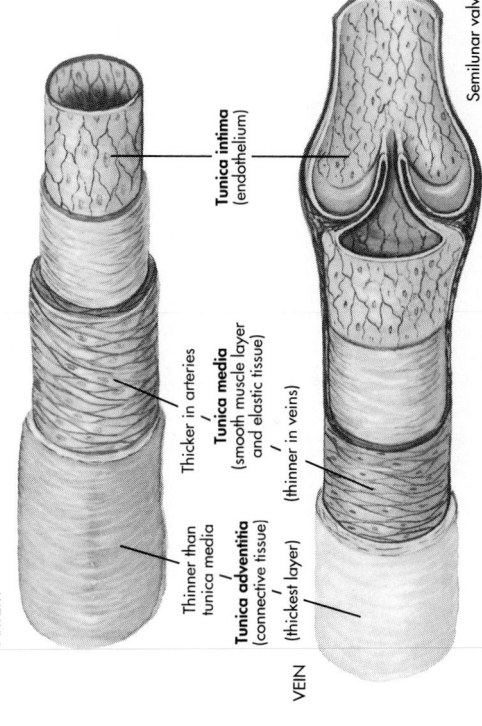

ARTERY

Thicker in arteries

Tunica media
(smooth muscle layer
and elastic tissue)

(thinner in veins)

Tunica intima
(endothelium)

Thinner than
tunica media

Tunica adventitia
(connective tissue)
(thickest layer)

VEIN

Semilunar valve

Structure of blood vessels.

STRUCTURE OF BLOOD VESSELS

Type of Vessel	Tunica Intima (Endothelium)	Tunica Media (Smooth Muscle; Elastic Connective Tissue)	Tunica Adventitia (Fibrous Connective Tissue)
ARTERIES	Smooth lining	Allows constriction and dilation of vessels; thicker than in veins; muscle innervated by autonomic fibers	Provides flexible support that resists collapse or injury; thicker than in veins; thinner than tunica media
VEINS	Smooth lining with semilunar valves to ensure one-way flow	Allows constriction and dilation of vessels; thinner than in arteries; muscle innervated by autonomic fibers	Provides flexible support that resists collapse or injury; thinner than in arteries; thicker than tunica media
CAPILLARIES	Makes up entire wall of capillary; thinness permits ease of transport across vessel wall	(Absent)	(Absent)

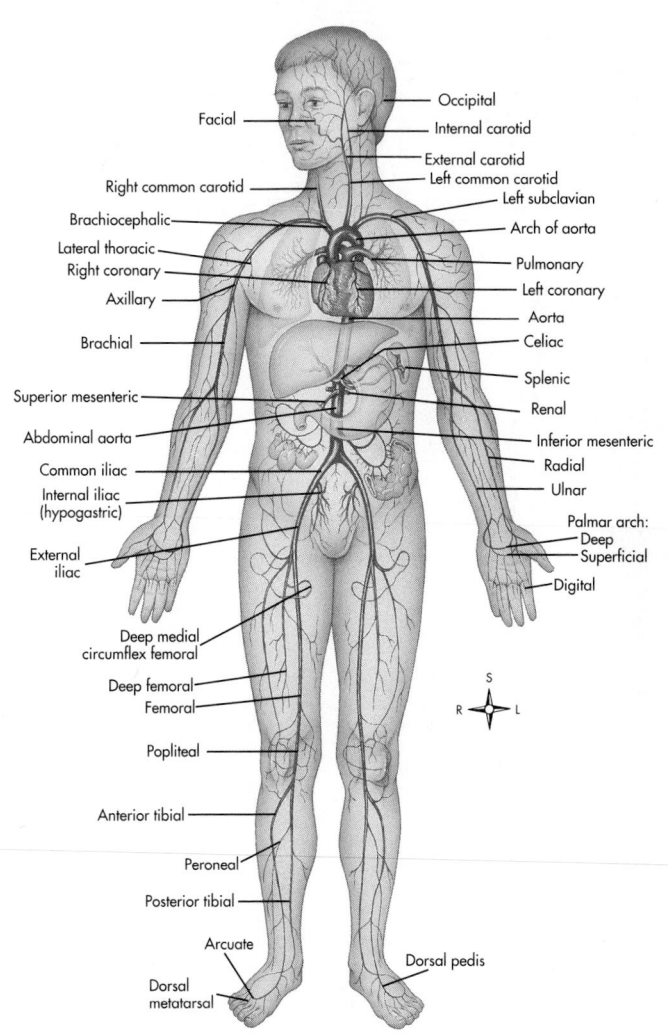

Principal arteries of the body.

Artery*	Region Supplied
ASCENDING AORTA	
Coronary arteries	Myocardium
ARCH OF AORTA	
Brachiocephalic (innominate)	Head and upper extremity
Right subclavian	Head, upper extremity
Right vertebral	Spinal cord, brain
Right axillary (continuation of subclavian)	Shoulder, chest, axillary region
Right brachial (continuation of axillary)	Arm and hand
Right radial	Lower arm and hand (lateral)
Right ulnar	Lower arm and hand (medial)
Superficial and deep palmar arches (formed by anastomosis of branches of radial and ulnar)	Hand and fingers
Digital	Fingers
Right common carotid	Head and neck
Right internal carotid	Brain, eye, forehead, nose
Right external carotid	Thyroid, tongue, tonsils, ear, etc.

Artery*	Region Supplied
DESCENDING ABDOMINAL AORTA	
Visceral branches	
Celiac artery (trunk)	Abdominal viscera
Left gastric	Stomach, esophagus
Common hepatic	Liver
Splenic	Spleen, pancreas, stomach
Superior mesenteric	Pancreas, small intestine, colon
Inferior mesenteric	Descending colon, rectum
Suprarenal	Adrenal (suprarenal) gland
Renal	Kidney
Ovarian	Ovary, uterine tube, ureter
Testicular	Testis, ureter
Parietal branches	Walls of abdomen
Inferior phrenic	Inferior surface of diaphragm, adrenal gland
Lumbar	Lumbar vertebrae and muscles of back
Median sacral	Lower vertebrae
Common iliac (formed by terminal branches of aorta)	**Pelvis, lower extremity**

Left subclavian — Head and upper extremity
- Left vertebral — Spinal cord, brain
- Left axillary (continuation of subclavian) — Shoulder, chest, axillary region
- Left brachial (continuation of axillary) — Arm and hand
  - Left radial — Lower arm and hand (lateral)
  - Left ulnar — Lower arm and hand (medial)
    - Superficial and deep palmar arches (formed by anastomosis of branches of radial and ulnar) — Hand and fingers
      - Digital — Fingers

Left common carotid — Head and neck
- Left internal carotid — Brain, eye, forehead, nose
- Left external carotid — Thyroid, tongue, tonsils, ear, etc.

DESCENDING THORACIC AORTA

Visceral branches — Thoracic viscera
- Bronchial — Lungs, bronchi
- Esophageal — Esophagus

Parietal branches — Thoracic walls
- Intercostal — Lateral thoracic walls (rib cage)
- Superior phrenic — Superior surface of diaphragm

External iliac
- Femoral (continuation of external iliac) — Thigh, leg, foot
- Popliteal (continuation of femoral) — Thigh, leg, foot
  - Anterior tibial — Leg, foot
  - Posterior tibial — Leg, foot
    - Plantar arch (formed by anastomosis of branches of anterior and posterior tibial arteries) — Leg, foot
      - Digital — Foot, toes; Toes

Internal iliac — Pelvis
- **Visceral branches** — Pelvic viscera
  - Middle rectal — Rectum
  - Vaginal — Vagina, uterus
  - Uterine — Uterus, vagina, uterine tube, ovary
- **Parietal branches** — Pelvic wall and external regions
  - Lateral sacral — Sacrum
  - Superior gluteal — Gluteal muscles
  - Obturator — Pubic region, hip joint, groin
  - Internal pudendal — Rectum, external genitals, floor of pelvis
  - Inferior gluteal — Lower gluteal region, coccyx, upper thigh

*Branches of each artery are indented below its name.

MAJOR SYSTEMIC VEINS

Vein*	Region Drained
SUPERIOR VENA CAVA	Head, neck, thorax, upper extremity
Brachiocephalic (innominate)	Head, neck, upper extremity
Internal jugular (continuation of sigmoid sinus)	Brain
Lingual	Tongue, mouth
Superior thyroid	Thyroid, deep face
Facial	Superficial face
Sigmoid sinus (continuation of transverse sinus/direct tributary of internal jugular)	Brain, meninges, skull
Superior and inferior petrosal sinuses	Anterior brain, skull
Cavernous sinus	Anterior brain, skull
Ophthalmic veins	Eye, orbit
Transverse sinus (direct tributary of sigmoid sinus)	Brain, meninges, skull
Occipital sinus	Inferior, central region of cranial cavity
Straight sinus	Central region of brain, meninges
Inferior sagittal sinus	Central region of brain, meninges
Superior sagittal (longitudinal) sinus	Superior region of cranial cavity
External jugular	Superficial, posterior head, neck

Vein*	Region Drained
INFERIOR VENA CAVA	Lower trunk and extremity
Phrenic	Diaphragm
Hepatic portal system	Upper abdominal viscera
Hepatic veins (continuations of liver venules and sinusoids and, ultimately, the hepatic portal vein)	Liver
Hepatic portal vein	Gastrointestinal organs, pancreas, spleen, gallbladder
Cystic	Gallbladder
Gastric	Stomach
Splenic	Spleen
Inferior mesenteric	Descending colon, rectum
Pancreatic	Pancreas
Superior mesenteric	Small intestine, most of colon
Gastroepiploic	Stomach
Renal	Kidneys
Suprarenal	Adrenal (suprarenal) gland
Left ovarian	Left ovary
Left testicular	Left testis
Left ascending lumbar (anastomoses with hemiazygos)	Left lumbar region
Right ovarian	Right ovary
Right testicular	Right testis

Vein	Region drained
Right ascending lumbar (anastomoses with azygos)	Right lumbar region
Common iliac (continuation of external iliac; common iliacs unite to form inferior vena cava)	Lower extremity
External iliac (continuation of femoral/direct tributary of common iliac)	Thigh, leg, foot
Femoral (continuation of popliteal/direct tributary of external iliac)	Thigh, leg, foot
Popliteal	Leg, foot
Posterior tibial	Deep posterior leg
Medial and lateral plantar	Sole of foot
Fibular (peroneal) (continuation of anterior tibial)	Lateral and anterior leg, foot
Anterior tibial	Anterior leg, foot
Dorsal veins of foot	Anterior (dorsal) foot, toes
Small (external, short) saphenous	Superficial posterior leg, lateral foot
Great (internal, long) saphenous	Superficial medial and anterior thigh, leg, foot
Dorsal veins of foot	Anterior (dorsal) foot, toes
Dorsal venous arch	Anterior (dorsal) foot, toes
Digital	Toes
Internal iliac	Pelvic region

Vein	Region drained
Subclavian (continuation of axillary/direct tributary of brachiocephalic)	Axilla, lower extremity
Axillary (continuation of basilic/direct tributary of subclavian)	Axilla, lower extremity
Cephalic	Lateral and lower arm, hand
Brachial	Deep arm
Radial	Deep lateral forearm
Ulnar	Deep medial forearm
Basilic (direct tributary of axillary)	Medial and lower arm, hand
Median cubital (basilic) (formed by anastomosis of cephalic and basilic)	Arm, hand
Deep and *superficial palmar* venous arches (formed by anastomosis of cephalic and basilic)	Hand
Digital	Fingers
Azygos (anastomoses with right ascending lumbar)	Right posterior wall of thorax and abdomen, esophagus, bronchi, pericardium, mediastinum
Hemiazygos (anastomoses with left renal)	Left inferior posterior wall of thorax and abdomen, esophagus, mediastinum
Accessory hemiazygos	Left superior posterior wall of thorax

*Tributaries of each vein are indented below its name; deep veins are printed in **bold type** and superficial veins are printed in *italic type*.

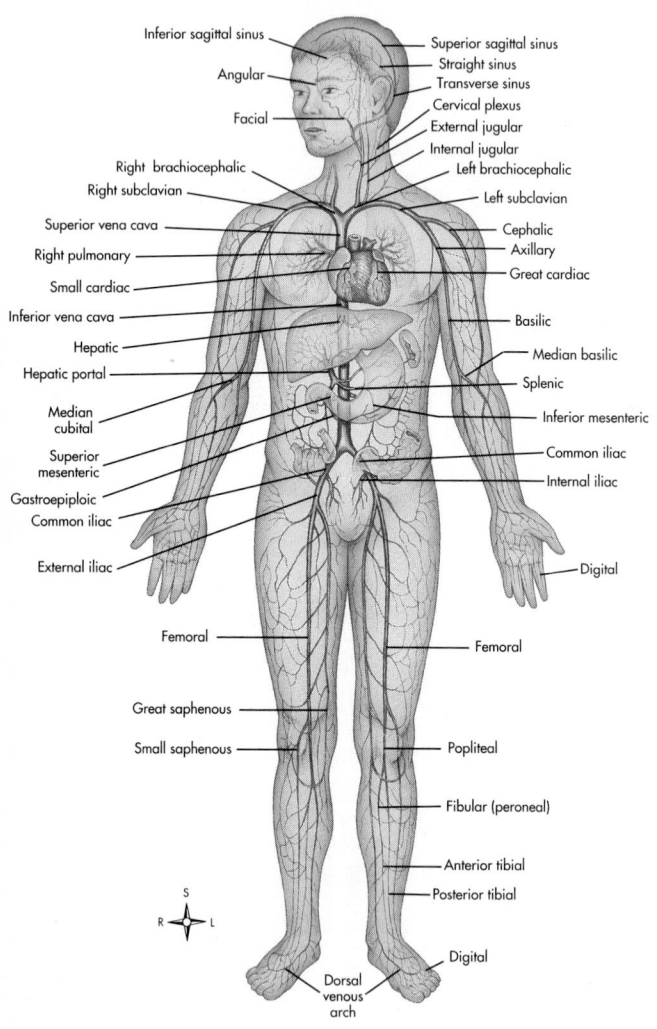

Principal veins of the body.

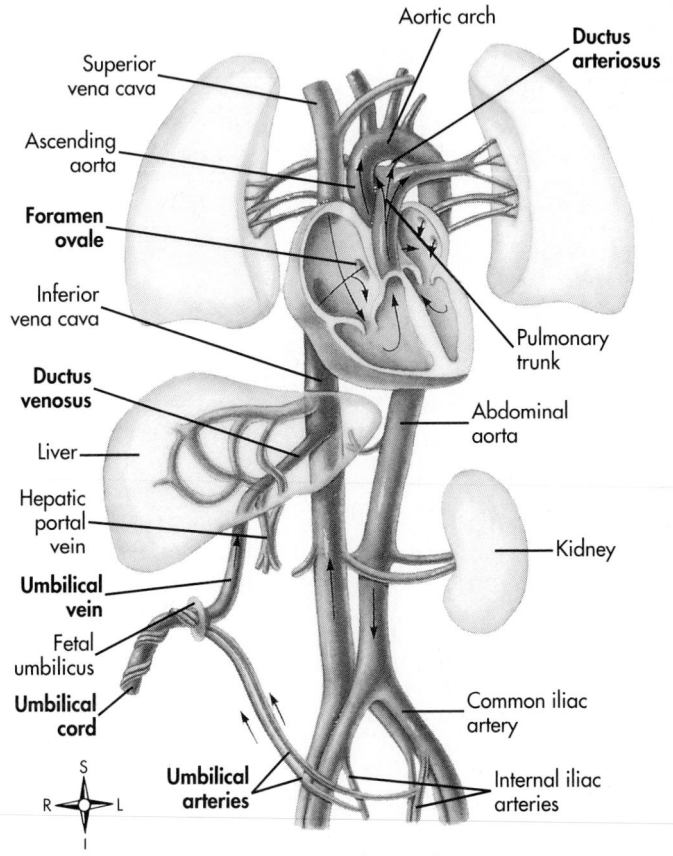

Plan of fetal circulation. Before birth the human circulatory system has several special features that adapt the body to life in the womb. These features (labeled in bold type) include: two umbilical arteries, one umbilical vein, ductus venosus, foramen ovale, and ductus arteriosus. The placenta, another essential feature of the fetal circulatory plan is not shown.

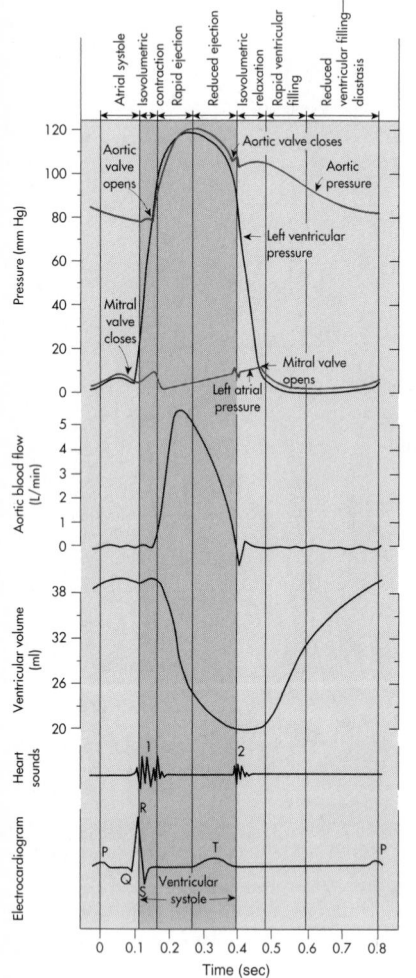

Composite chart of heart function. This chart is a composite of several different diagrams of heart function: cardiac pumping cycle, blood pressure, blood flow, volume, heart sounds, and ECG.

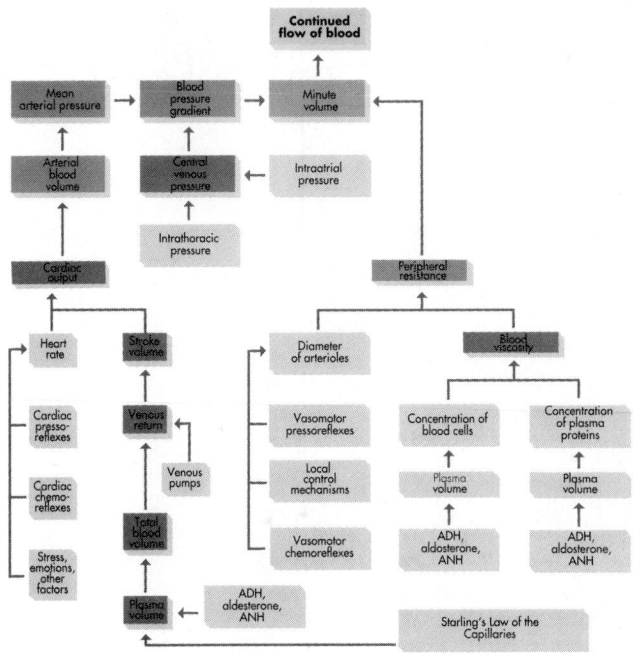

Factors that influence the flow of blood. The flow of blood, expressed as volume of blood flowing per minute (or minute volume), is determined by a great variety of factors. This chart shows only some of the major factors that influence blood flow. Notice that some factors appear more than once in the chart, indicating that they can influence blood flow in several different ways.

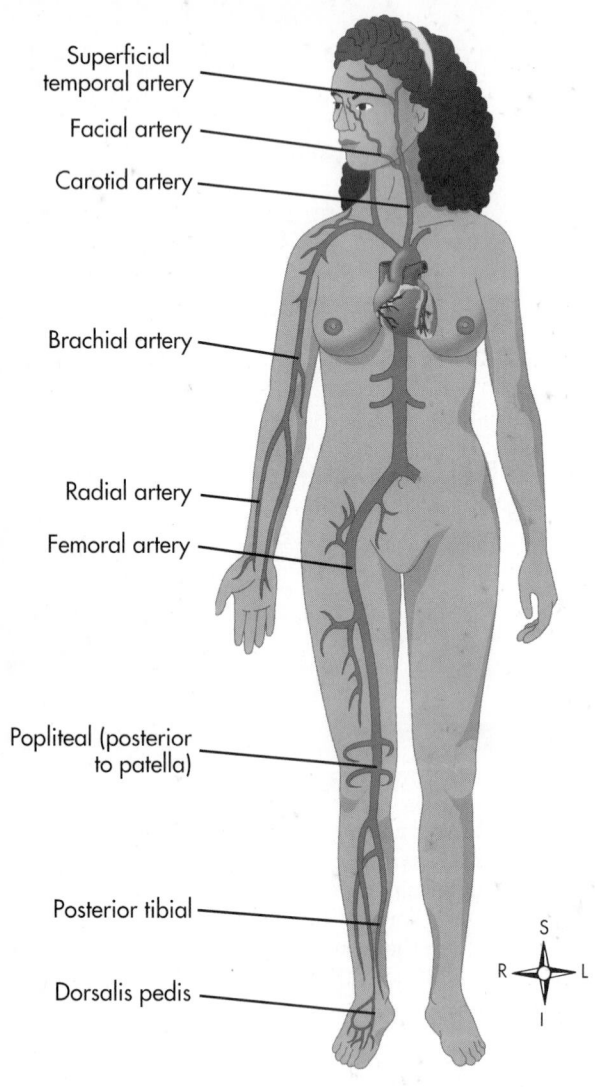

Superficial temporal artery
Facial artery
Carotid artery
Brachial artery
Radial artery
Femoral artery
Popliteal (posterior to patella)
Posterior tibial
Dorsalis pedis

S
R — L
I

Pulse points. Each pulse point is named after the artery with which it is associated.

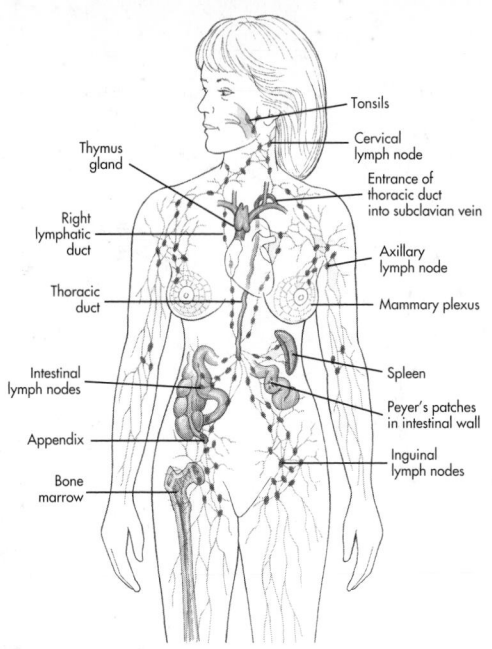

Location of principal lymphatic system organs.

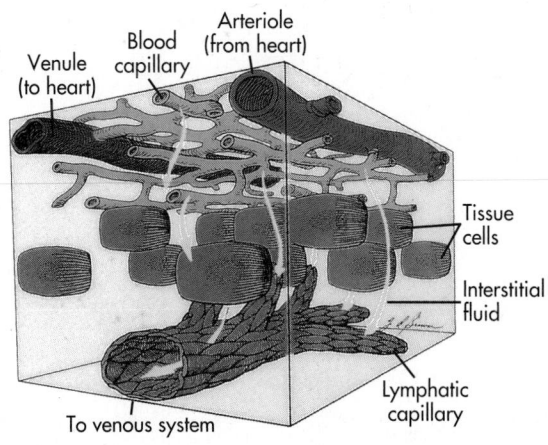

Role of the lymphatic system in fluid balance.

MECHANISMS OF NONSPECIFIC DEFENSE

Mechanism	Description
SPECIES RESISTANCE	Genetic characteristics of the human species protect the body from certain pathogens
MECHANICAL AND CHEMICAL BARRIERS	
Skin and mucosa	Forms a continuous wall that separates the internal environment from the external environment, preventing the entry of pathogens
Secretions	Secretions such as sebum, mucus, and enzymes chemically inhibit the activity of pathogens
INFLAMMATION	The inflammatory response isolates the pathogens and stimulates the speedy arrival of large numbers of immune cells
PHAGOCYTOSIS	Ingestion and destruction of pathogens by phagocytic cells
Neutrophils	Granular leukocytes that are usually the first phagocytic cell to arrive at the scene of an inflammatory response
Macrophages	Monocytes that have enlarged to become giant phagocytic cells capable of consuming many pathogens; often called by other, more specific, names when found in specific tissues of the body
NATURAL KILLER (NK) CELLS	Group of lymphocytes that kill many different types of cancer cells and virus-infected cells
INTERFERON	Protein produced by cells after they become infected by a virus; inhibits the spread or further development of a viral infection
COMPLEMENT	Group of plasma proteins (inactive enzymes) that produce a cascade of chemical reactions that ultimately causes lysis (rupture) of a foreign cell; the complement cascade can be triggered by specific or nonspecific immune mechanisms

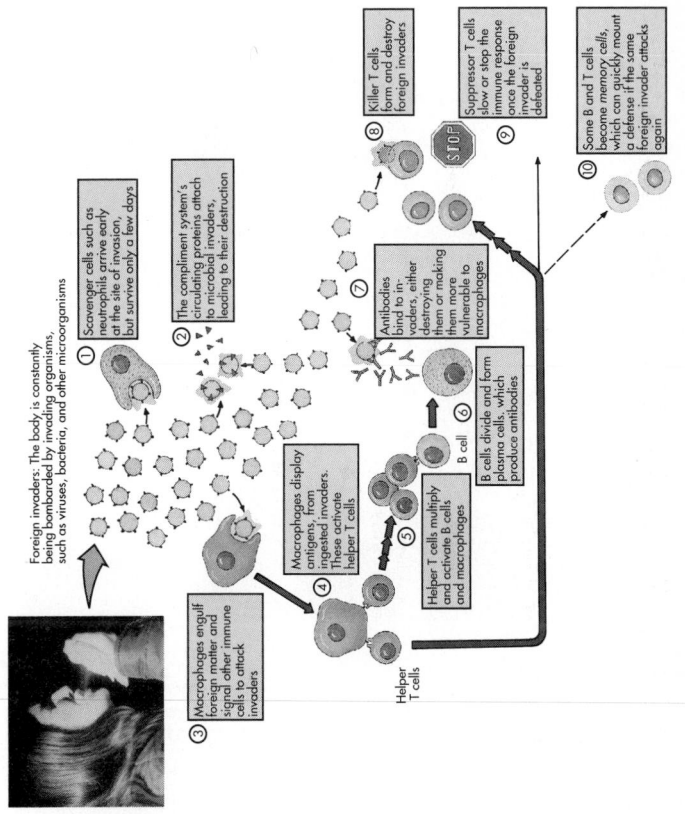

Foreign invaders: The body is constantly being bombarded by invading organisms, such as viruses, bacteria, and other microorganisms

① Scavenger cells such as neutrophils arrive early at the site of invasion but survive only a few days

② The compliment system's circulating proteins attach to microbial invaders, leading to their destruction

③ Macrophages engulf foreign matter and signal other immune cells to attack invaders

④ Macrophages display antigens, from ingested invaders. These activate helper T cells

Helper T cells

⑤ Helper T cells multiply and activate B cells and macrophages

⑥ B cells divide and form plasma cells, which produce antibodies

B cell

⑦ Antibodies bind to invaders, either destroying them or making them more vulnerable to macrophages

⑧ Killer T cells form and destroy foreign invaders

⑨ Suppressor T cells slow or stop the immune response once the foreign invader is defeated

STOP

⑩ Some B and T cells become memory cells, which can quickly mount a defense if the same foreign invader attacks again

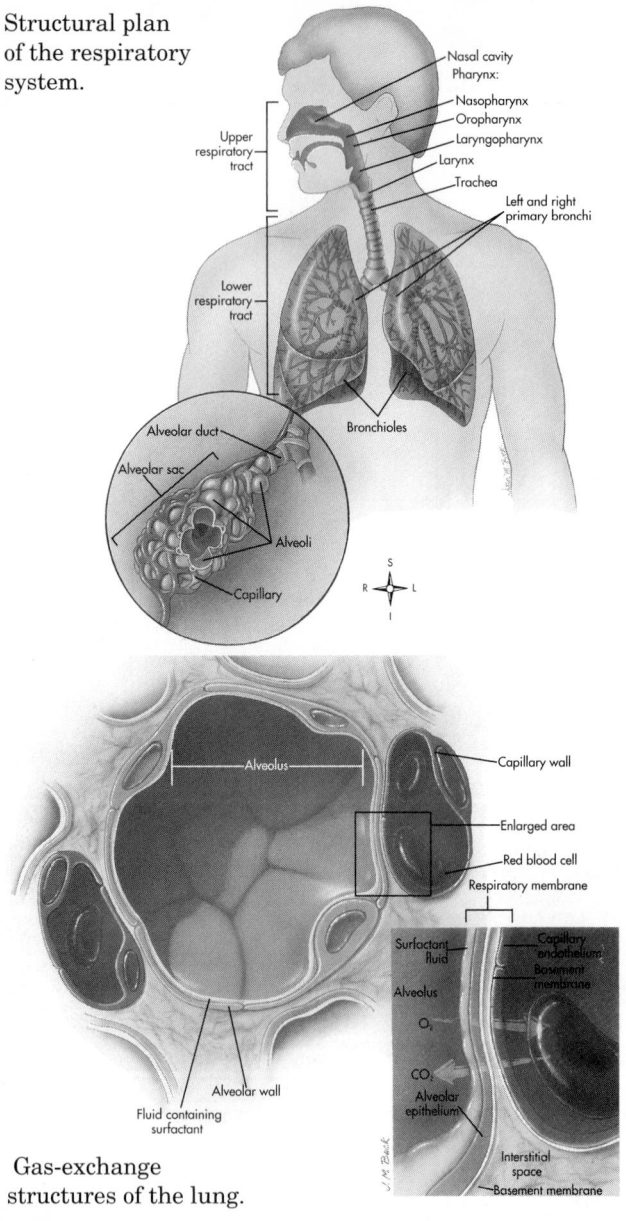

Structural plan of the respiratory system.

- Nasal cavity
- Pharynx:
  - Nasopharynx
  - Oropharynx
  - Laryngopharynx
- Larynx
- Trachea
- Left and right primary bronchi

Upper respiratory tract

Lower respiratory tract

Bronchioles

Alveolar duct

Alveolar sac

Alveoli

Capillary

Gas-exchange structures of the lung.

Alveolus

Capillary wall

Enlarged area

Red blood cell

Respiratory membrane

Alveolar wall

Fluid containing surfactant

Surfactant fluid

Alveolus

O_2

CO_2

Alveolar epithelium

Capillary endothelium

Basement membrane

Interstitial space

Basement membrane

J.M. Beck

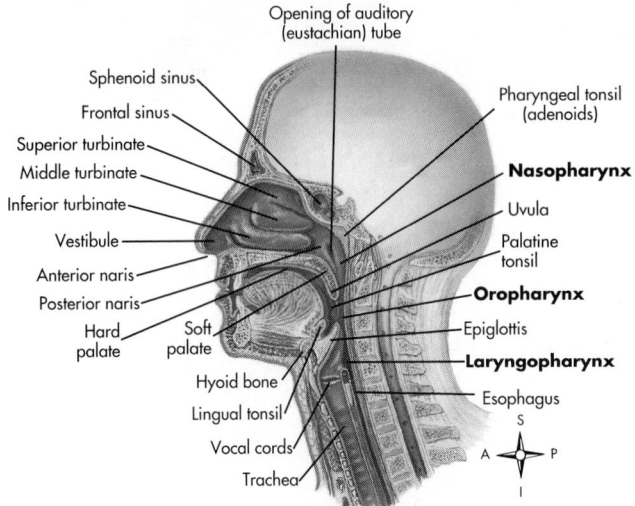

Midsagittal section of the head.

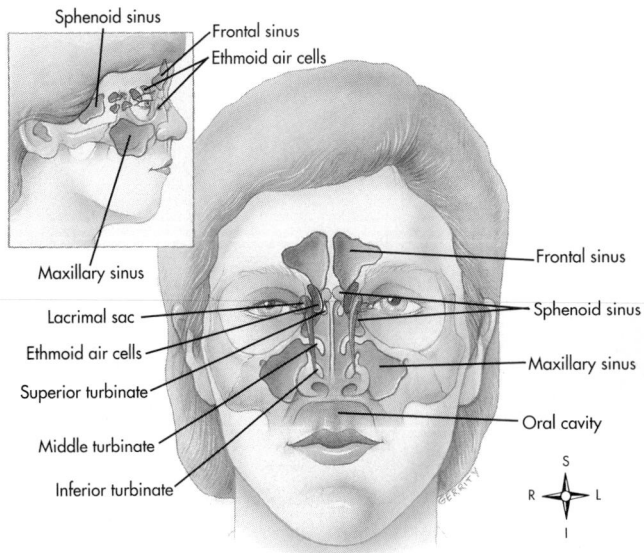

The paranasal sinuses.

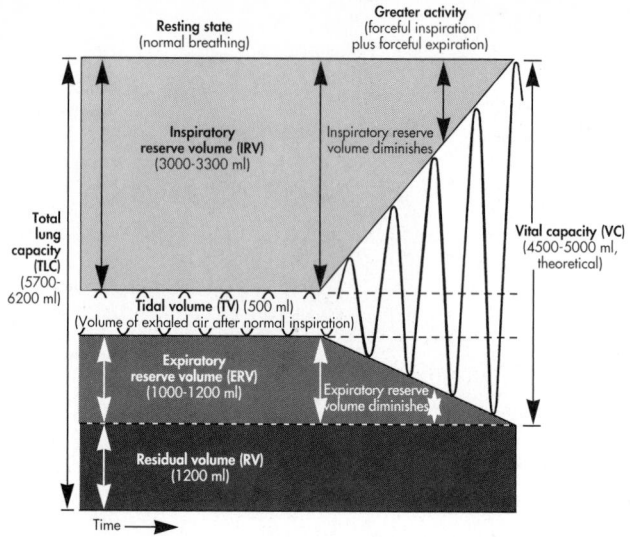

Spirogram.

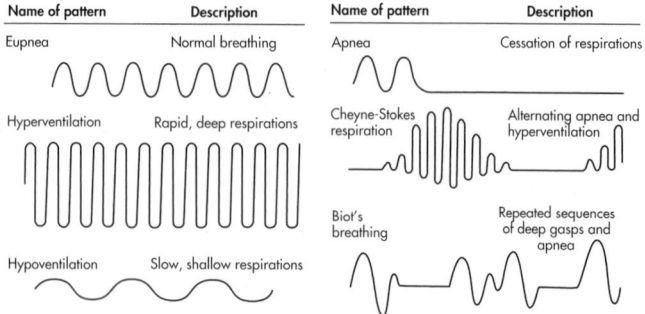

Breathing patterns.

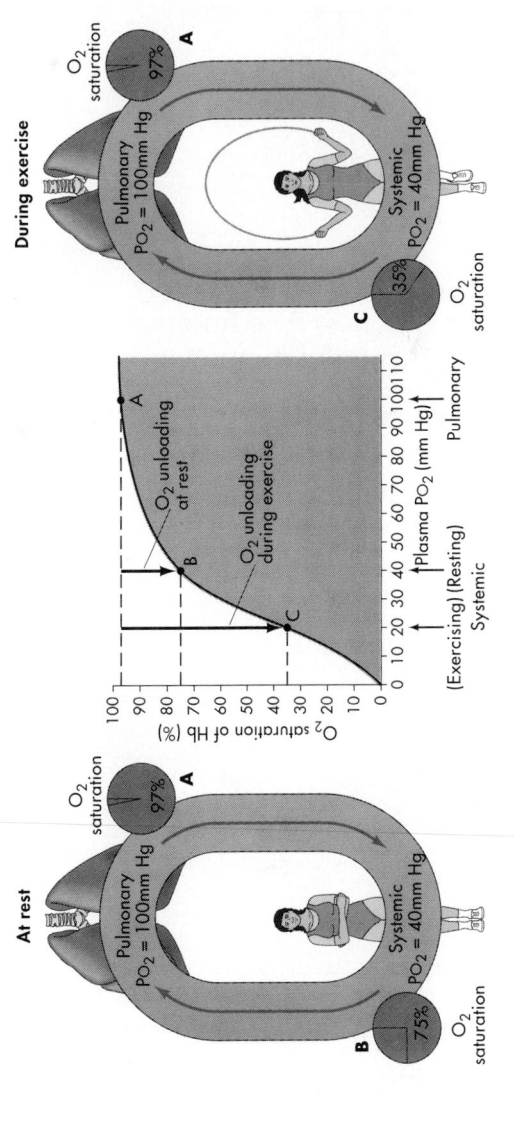

O_2 unloading at rest and during exercise. At rest, fully saturated Hb unloads almost 25% of its O_2 load when it reaches the low PO_2 (40 mm Hg) environment in systemic tissues. During exercise, the tissue PO_2 is even lower (20 mm Hg), causing fully saturated Hb to unload about 70% of its O_2 load. As you can see, a slight drop in tissue PO_2 causes a large increase in O_2 unloading.

115

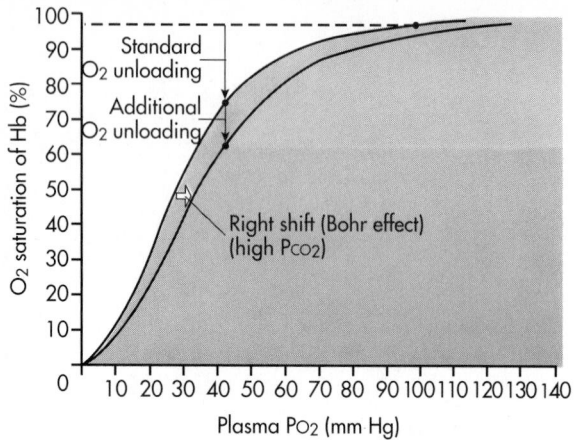

Right shift of oxyhemoglobin dissociation curve.

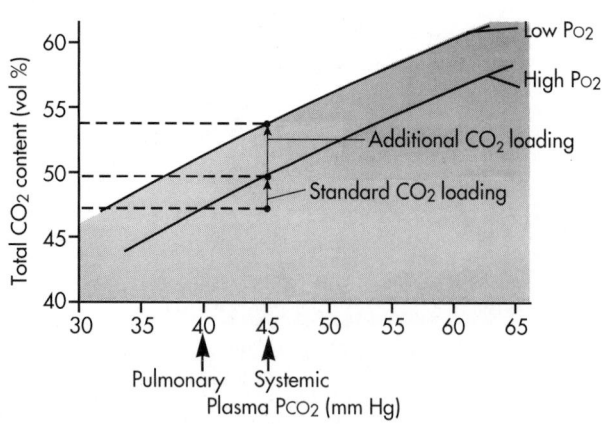

Carbon dioxide dissociation curve.

Parotid gland
Submandibular gland
Pharynx
Trachea
Esophagus
Diaphragm
Liver
Hepatic flexure
Transverse colon
Ascending colon
Ileum
Cecum
Region of ileocecal valve
Vermiform appendix
Tongue
Sublingual gland
Larynx
Hepatic bile duct
Liver
Cystic duct
Gallbladder
Duodenum
Pancreas
Spleen
Spleen
Splenic flexure
Stomach
Descending colon
Sigmoid colon
Rectum
Anal canal
S
R
L
I

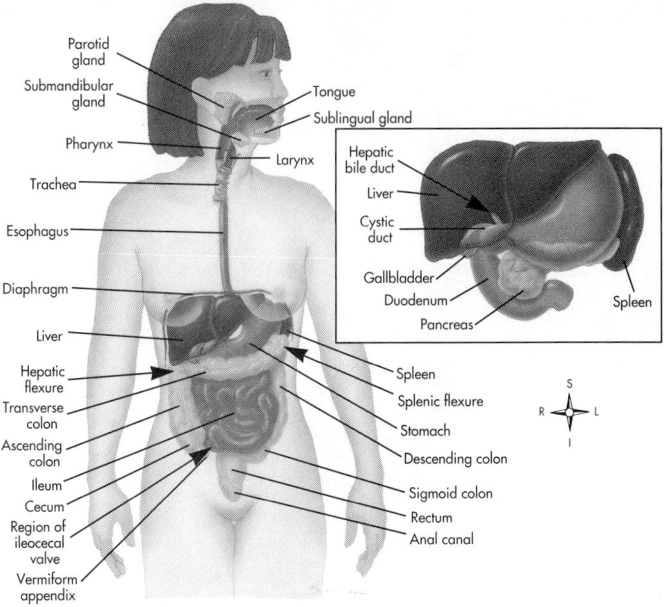

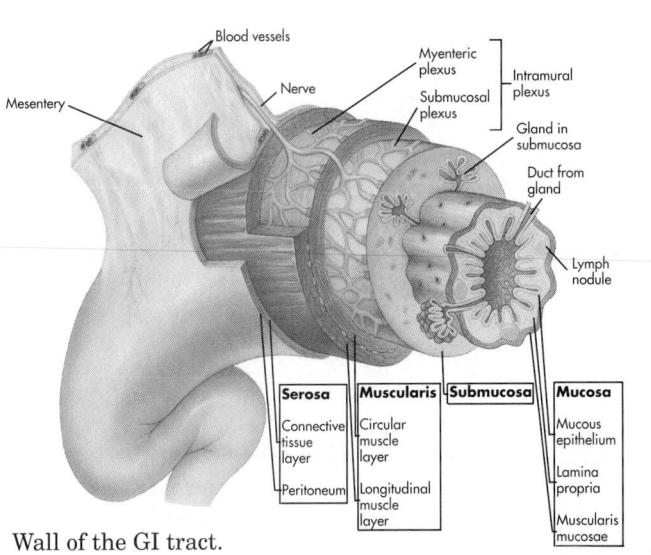

Location of digestive organs.

Blood vessels
Mesentery
Nerve
Myenteric plexus
Submucosal plexus
Intramural plexus
Gland in submucosa
Duct from gland
Lymph nodule
Serosa
Connective tissue layer
Peritoneum
Muscularis
Circular muscle layer
Longitudinal muscle layer
Submucosa
Mucosa
Mucous epithelium
Lamina propria
Muscularis mucosae

Wall of the GI tract.

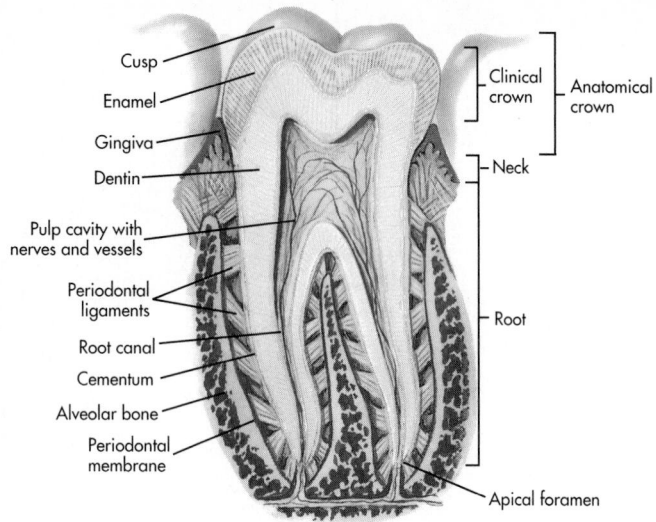

Typical tooth.

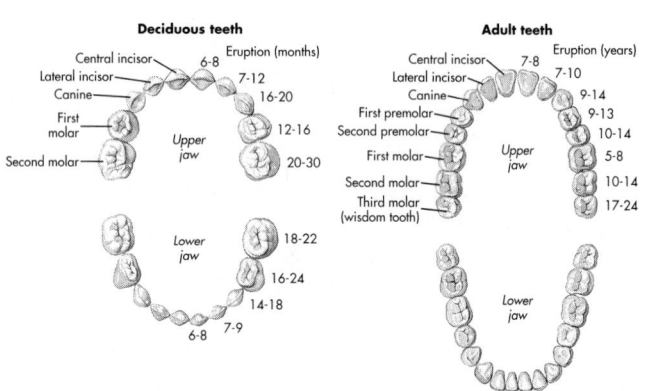

Deciduous (baby) teeth and adult teeth. In the deciduous set, there are no premolars and only two pairs of molars in each jaw. Generally the lower teeth erupt before the corresponding upper teeth.

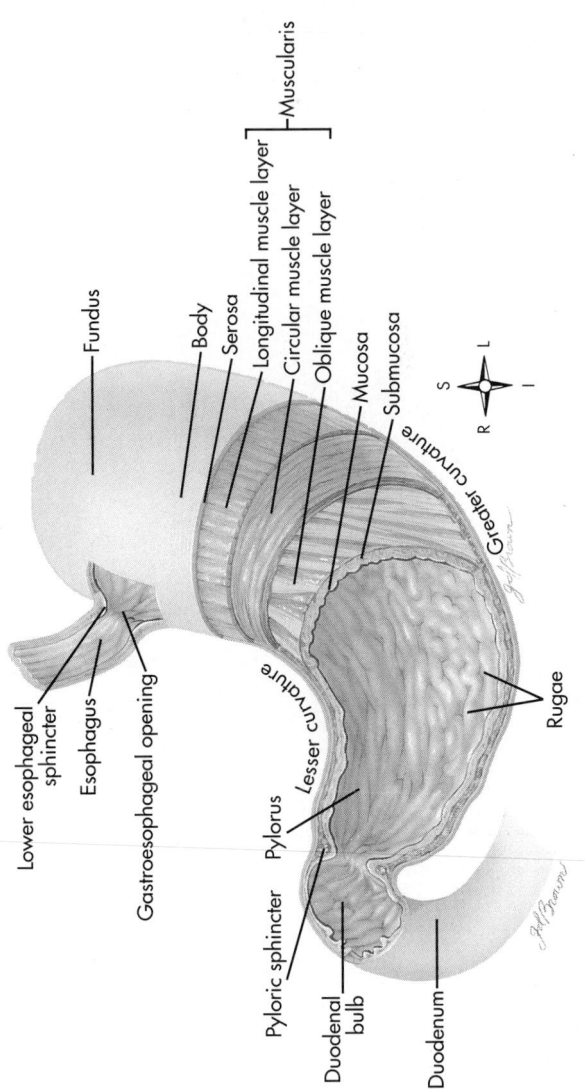

Fundus

Body

Serosa

Longitudinal muscle layer

Circular muscle layer — Muscularis

Oblique muscle layer

Mucosa

Submucosa

Greater curvature

Lower esophageal sphincter

Esophagus

Gastroesophageal opening

Lesser curvature

Rugae

Pylorus

Pyloric sphincter

Duodenal bulb

Duodenum

S

R   L

I

Stomach.

119

Divisions of the large intestine.

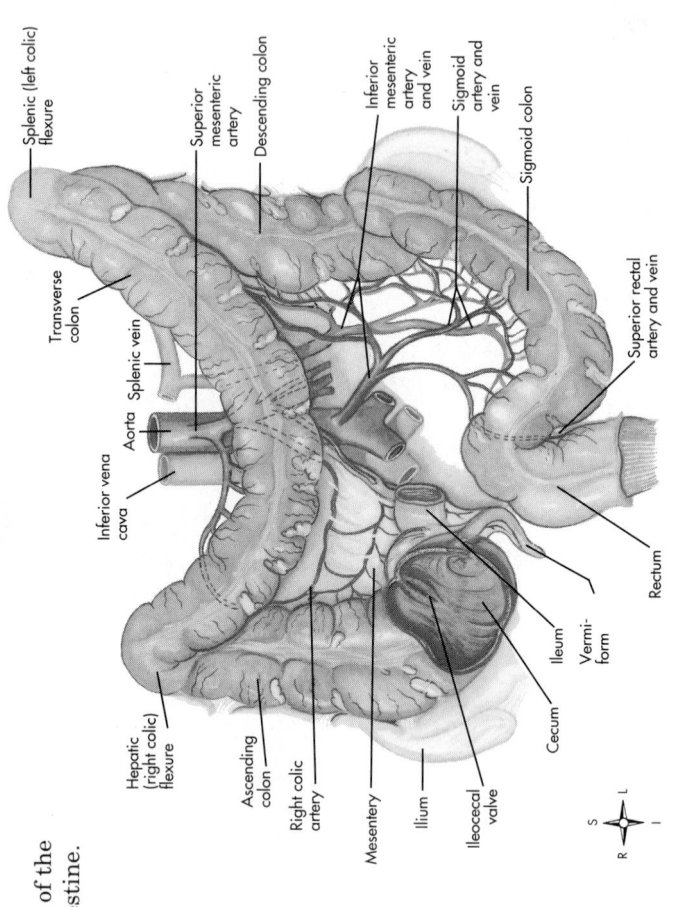

Splenic (left colic) flexure

Superior mesenteric artery

Descending colon

Inferior mesenteric artery and vein

Sigmoid artery and vein

Sigmoid colon

Transverse colon

Splenic vein

Aorta

Inferior vena cava

Superior rectal artery and vein

Rectum

Hepatic (right colic) flexure

Ascending colon

Right colic artery

Mesentery

Ilium

Ileocecal valve

Cecum

Ileum

Vermiform

S
R — L
I

PRIMARY MECHANISMS OF THE DIGESTIVE SYSTEM

Mechanism	Description
Ingestion	Process of taking food into the mouth, starting it on its journey through the digestive tract
Digestion	A group of processes that break complex nutrients into simpler ones, thus facilitating their absorption; *mechanical digestion* physically breaks large chunks into small bits; *chemical digestion* breaks molecules apart
Motility	Movement by the muscular components of the digestive tube, including processes of mechanical digestion; examples include *peristalsis* and *segmentation*
Secretion	Release of digestive juices (containing enzymes, acids, bases, mucus, bile, or other products that facilitate digestion); some digestive organs also secrete endocrine hormones that regulate digestion or metabolism of nutrients
Absorption	Movement of digested nutrients through the GI mucosa and into the internal environment
Elimination	Excretion of the residues of the digestive process (feces) from the rectum, through the anus; defecation

PROCESSES OF MECHANICAL DIGESTION

Organ	Mechanical Process	Nature of Process
Mouth (teeth and tongue)	Mastication	Chewing movements—reduce size of food particles and mix them with saliva
	Deglutition	Swallowing—movement of food from mouth to stomach
Pharynx	Deglutition	
Esophagus	Deglutition	
	Peristalsis	Rippling movements that squeeze food downward in tract; constricted ring forms first in one section, the next, etc., causing waves of contraction to spread along entire canal
Stomach	Churning	Forward and backward movement of gastric contents, mixing food with gastric juices to form chyme
	Peristalsis	Wave starting in body of stomach about three times per minute and sweeping toward closed pyloric sphincter; at intervals, strong peristaltic waves press chyme past sphincter into duodenum
Small intestine	Segmentation (mixing contractions)	Forward and backward movement within segment of intestine; purpose, to mix food and digestive juices thoroughly and to bring all digested food in contact with intestinal mucosa to facilitate absorption; purpose of peristalsis, on the other hand, to propel intestinal contents along digestive tract
	Peristalsis	
Large intestine		
Colon	Segmentation	Churning movements within haustral sacs
	Peristalsis	
Descending colon	Mass peristalsis	Entire contents moved into sigmoid colon and rectum; occurs three or four times a day, usually after a meal
Rectum	Defecation	Emptying of rectum, so-called bowel movement

CHEMICAL DIGESTION

Digestive Juices and Enzymes	Substance Digested (or Hydrolyzed)	Resulting Product*
Saliva		
Amylase	Starch (polysaccharide)	Maltose (a double sugar, or disaccharide)
Gastric Juice		
Protease (pepsin) plus hydrochloric acid	Proteins	Partially digested proteins
Pancreatic Juice		
Proteases (e.g., trypsin)[†]	Proteins (intact or partially digested)	Peptides and **amino acids**
Lipases	Fats emulsified by bile	**Fatty acids, monoglycerides, and glycerol**
Amylase	Starch	Maltose
Intestinal Enzymes[‡]		
Peptidases	Peptides	**Amino acids**
Sucrase	Sucrose (cane sugar)	**Glucose and fructose[§]** (simple sugars, or monosaccharides)
Lactase	Lactose (milk sugar)	**Glucose and galactose** (simple sugars)
Maltase	Maltose (malt sugar)	**Glucose**

*Substances in boldface type are end products of digestion (that is, completely digested nutrients ready for absorption).

[†]Secreted in inactive form (trypsinogen); activated by enterokinase, an enzyme in the intestinal brush border.

[‡]Brush-border enzymes.

[§]Glucose is also called *dextrose;* fructose is also called *levulose.*

ACTIONS OF SOME DIGESTIVE HORMONES SUMMARIZED

Hormone	Source	Action
Gastrin	Formed by gastric mucosa in presence of partially digested proteins, when stimulated by the vagus nerve, or when the stomach is stretched	Stimulate secretion of gastric juice rich in pepsin and hydrochloric acid
Gastric inhibitory peptide (GIP)	Formed by intestinal mucosa in presence of fats and perhaps other nutrients	Inhibits gastric secretion and motility
Secretin	Formed by intestinal mucosa in presence of acid, partially digested proteins, and fats	Inhibits gastric secretion; stimulates secretion of pancreatic juice low in enzymes and high in alkalinity (bicarbonate); stimulates ejection of bile by the gallbladder
Cholecystokinin-pancreozymin (CCK)	Formed by intestinal mucosa in presence of fats, partially digested proteins, acids	Stimulates ejection of bile from gallbladder and secretion of pancreatic juice high in enzymes; opposes the action of gastrin, raising the pH of gastric juice

FOOD ABSORPTION

Form Absorbed	Structures into which Absorbed	Circulation
Protein—as amino acids	Blood in intestinal capillaries	Portal vein, liver, hepatic vein, inferior vena cava to heart, etc.
Perhaps minute quantities of some short-chain polypeptides and whole proteins absorbed, for example, some antibodies		
Carbohydrates—as simple sugars	Same as amino acids	Same as amino acids
Fats		
Glycerol and monoglycerides	Lymph in intestinal lacteals	During absorption, that is, while in epithelial cells of intestinal mucosa, glycerol and fatty acids recombine to form microscopic packages of fats (chylomicrons); lymphatics carry them by way of thoracic duct to left subclavian vein, superior vena cava, heart, etc.; some fats transported by blood in form of phospholipids or cholesterol esters
Fatty acids combine with bile salts to form water soluble substance	Lymph in intestinal lacteals	
Some finely emulsified, undigested fats absorbed	Small fraction enters intestinal blood capillaries	

125

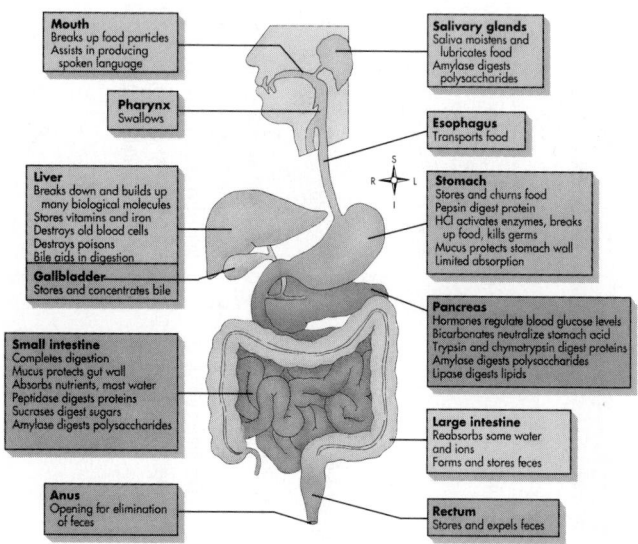

Mouth
Breaks up food particles
Assists in producing
spoken language

Pharynx
Swallows

Liver
Breaks down and builds up
many biological molecules
Stores vitamins and iron
Destroys old blood cells
Destroys poisons
Bile aids in digestion

Gallbladder
Stores and concentrates bile

Small intestine
Completes digestion
Mucus protects gut wall
Absorbs nutrients, most water
Peptidase digests proteins
Sucrases digest sugars
Amylase digests polysaccharides

Anus
Opening for elimination
of feces

Salivary glands
Saliva moistens and
lubricates food
Amylase digests
polysaccharides

Esophagus
Transports food

Stomach
Stores and churns food
Pepsin digest protein
HCl activates enzymes, breaks
up food, kills germs
Mucus protects stomach wall
Limited absorption

Pancreas
Hormones regulate blood glucose levels
Bicarbonates neutralize stomach acid
Trypsin and chymotrypsin digest proteins
Amylase digests polysaccharides
Lipase digests lipids

Large intestine
Reabsorbs some water
and ions
Forms and stores feces

Rectum
Stores and expels feces

Summary of digestive organ functions.

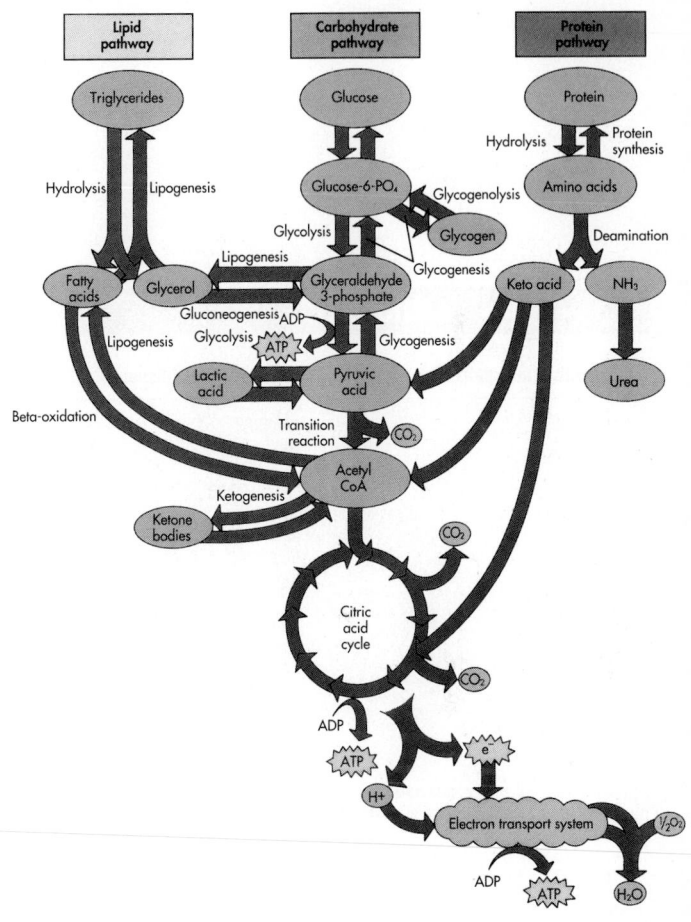

Summary of metabolism. Notice the central role played by the citric acid cycle and electron transport system. Notice also how different molecules can be converted to forms that may enter other pathways.

AMINO ACIDS

Essential (indispensible)	Nonessential (dispensable)
Histidine*	Alanine
Isoleucine	Arginine
Leucine	Asparagine
Lysine	Aspartic acid
Methionine	Cysteine
Phenylalanine	Glutamic acid
Threonine	Glutamine
Tryptophan	Glycine
Valine	Proline
	Serine
	Tyrosine[†]

*Essential in infants and, perhaps, adult males.
†Can be synthesized from phenylalanine, so is nonessential if phenylalanine is in the diet.

METABOLISM

Nutrient	Anabolism	Catabolism
Carbohydrates	Temporary excess changed into glycogen by liver cells in presence of insulin; stored in liver and skeletal muscles until needed and then changed back to glucose	Oxidized, in presence of insulin, to yield energy (4.1 kcal per g) and wastes (carbon dioxide and water) $$C_6H_{12}O_6 \; + \; 6\,O_2 \; \longrightarrow \; \text{Energy} \; + \; 6\,CO_2 \; + \; 6\,H_2O$$
	True excess beyond body's energy requirements converted into adipose tissue; stored in various fat depots of body	
Fats	Built into adipose tissue; stored in fat depots of body	Fatty acids $\;\downarrow\;$ (beta-oxidation) Acetyl-CoA $\Longleftrightarrow$ Ketones $\;\downarrow\;$ (tissues; citric acid cycle) Energy (9.3 kcal per g) $+ \; CO_2 \; + \; H_2O$ Glycerol $\;\downarrow\;$ (glycolysis) Acetyl-CoA
Proteins	Synthesized into tissue proteins, blood proteins, enzymes, hormones, etc.	Deaminated by liver, forming ammonia (which is converted to urea) and keto acids (which are either oxidized or changed to glucose or fat)

MAJOR VITAMINS

Vitamin	Dietary Source	Functions	Symptoms of Deficiency
Vitamin A	Green and yellow vegetables, dairy products, and liver	Maintains epithelial tissue and produces visual pigments	Night blindness and flaking skin
B-complex vitamins			
B_1 (thiamine)	Grains, meat, and legumes	Helps enzymes in the critic acid cycle	Nerve problems (beriberi), heart muscle weakness, and edema
B_2 (riboflavin)	Green vegetables, organ meats, eggs, and dairy products	Aids enzymes in the critic acid cycle	Inflammation of skin and eyes
B_3 (niacin)	Meat and grains	Helps enzymes in the critic acid cycle	Pellagra (scaly dermatitis and mental disturbances) and nervous disorders
B_5 (pantothenic acid)	Organ meat, eggs, and liver	Aids enzymes that connect fat and carbohydrate metabolism	Loss of coordination (rare)
B_6 (pyridoxine)	Vegetables, meats, and grains	Helps enzymes that catabolize amino acids	Convulsions, irritability, and anemia
B_9 (folic acid)	Vegetables	Aids enzymes in amino acid catabolism and blood production	Digestive disorders and anemia
B_{12} (cyanocobalamin)	Meat and dairy products	Involved in blood production and other processes	Pernicious anemia
Biotin (vitamin H)	Vegetables, meat, and eggs	Helps enzymes in amino acid catabolism and fat and glycogen synthesis	Mental and muscle problems (rare)
Vitamin C (ascorbic acid)	Fruits and green vegetables	Helps in manufacture of collagen fibers; antioxidant	Scurvy and degeneration of skin, bone, and blood vessels
Vitamin D (calciferol)	Dairy products and fish liver oil	Aids in calcium absorption	Rickets and skeletal deformity
Vitamin E (tocopherol)	Green vegetables and seeds	Protects cell membranes from being destroyed	Muscle and reproductive disorders (rare)

MAJOR MINERALS

Mineral	Dietary Source	Functions	Symptoms of Deficiency
Calcium (Ca)	Dairy products, legumes, and vegetables	Helps blood clotting, bone formation, and nerve and muscle function	Bone degeneration and nerve and muscle malfunction
Chlorine (Cl)	Salty foods	Aids in stomach acid production and acid-base balance	Acid-base imbalance
Cobalt (Co)	Meat	Helps vitamin B_{12} in blood cell production	Pernicious anemia
Copper (Cu)	Seafood, organ meats, and legumes	Involved in extracting energy from the citric acid cycle and in blood production	Fatigue and anemia
Iodine (I)	Seafood and iodized salt	Required for thyroid hormone synthesis	Goiter (thyroid enlargement) and decrease of metabolic rate
Iron (Fe)	Meat, eggs, vegetables, and legumes	Involved in extracting energy from the citric acid cycle and in blood production	Fatigue and anemia
Magnesium (Mg)	Vegetables and grains	Helps many enzymes	Nerve disorders, blood vessel dilation, and heart rhythm problems
Manganese (Mn)	Vegetables, legumes, and grains	Helps many enzymes	Muscle and nerve disorders
Phosphorus (P)	Dairy products and meat	Aids in bone formation and is used to make ATP, DNA, RNA, and phospholipids	Bone degeneration and metabolic problems
Potassium (K)	Seafood, milk, fruit, and meats	Helps muscle and nerve function	Muscle weakness, heart problems, and nerve problems
Sodium (Na)	Salty foods	Aids in muscle and nerve function and fluid balance	Weakness and digestive upset
Zinc (Zn)	Many foods	Helps many enzymes	Inadequate growth

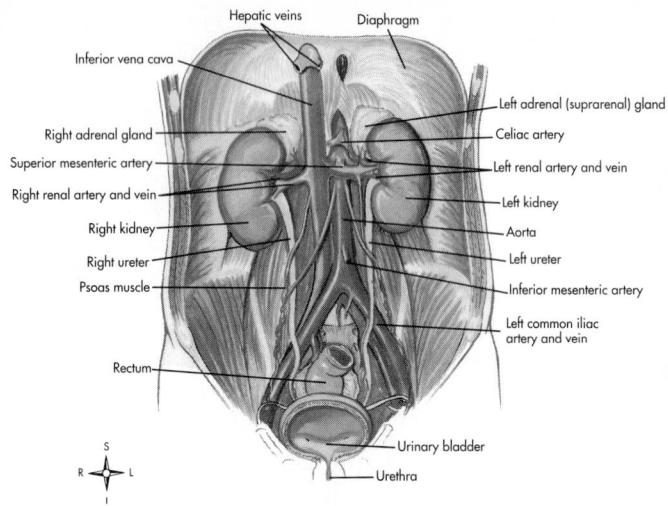

Location of urinary system organs.

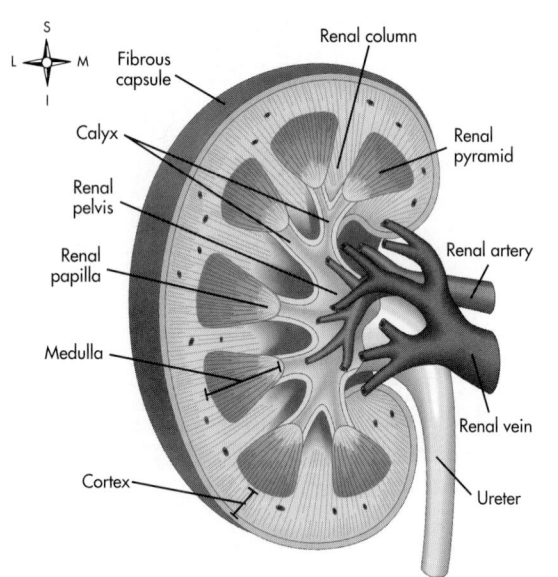

Kidney (frontal section).

132

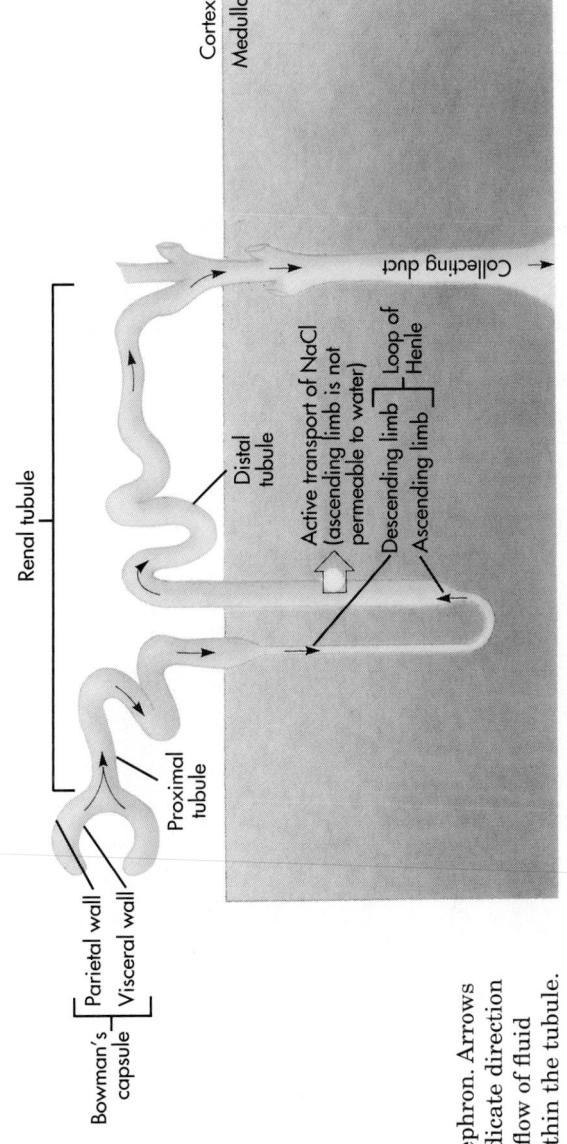

Nephron. Arrows indicate direction of flow of fluid within the tubule.

Labels within the figure:

Cortex

Medulla

Collecting duct

Distal tubule

Active transport of NaCl (ascending limb is not permeable to water)

Descending limb
Ascending limb
} Loop of Henle

Renal tubule

Proximal tubule

Parietal wall
Visceral wall
} Bowman's capsule

SUMMARY OF NEPHRON FUNCTION

Part of Nephron	Function	Substance Moved
Renal corpuscle	Filtration (passive)	Water Smaller solute particles (ions, glucose, etc.)
Proximal tubule	Reabsorption (active)	Active transport: Na+ Cotransport: glucose and amino acids
	Reabsorption (passive)	Diffusion: Cl−, PO4=, urea, other solutes Osmosis: water
Loop of Henle Descending limb	Reabsorption (passive)	Osmosis: water
	Secretion (passive)	Diffusion: urea
Ascending limb	Reabsorption (active)	Active transport: Na+
	Reabsorption (passive)	Diffusion: Cl−
Distal tubule	Reabsorption (active)	Active transport: Na+
	Reabsorption (passive)	Diffusion: Cl−, other anions Osmosis: water (only in presence of ADH)
	Secretion (passive)	Diffusion: ammonia
	Secretion (active)	Active transport: K+, H+, some drugs
Collecting duct	Reabsorption (active)	Active transport: Na+
	Reabsorption (passive)	Diffusion: urea Osmosis: water (only in presence of ADH)
	Secretion (passive)	Diffusion: ammonia
	Secretion (active)	Active transport: K+, H+, some drugs

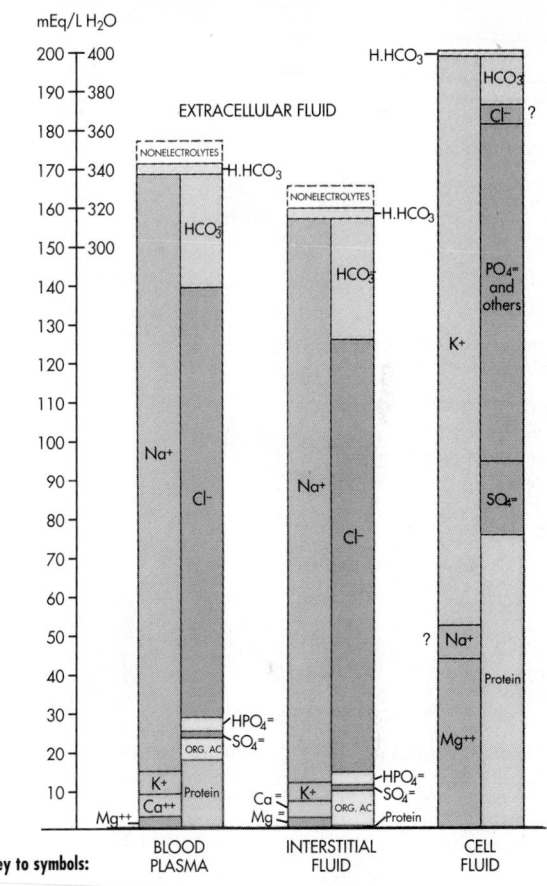

mEq/L H$_2$O

*Key to symbols:

Na$^+$	Sodium
K$^+$	Potassium
Mg^{++}	Magnesium
Ca^{++}	Calcium
Cl$^-$	Chloride
SO$_4$$^=$	Sulfate
HCO$_3$$^=$	Bicarbonate
CHPO$_4$$^=$	Phosphate
H.HCO$_3$	Carbonic acid

Chief chemical constituents of three fluid compartments. The column of figures at the left (200, 190, 180, ect.) indicates amounts of cations or anions. The figures on the right (400, 380, 360, etc.) indicate the sum of cations and anions.

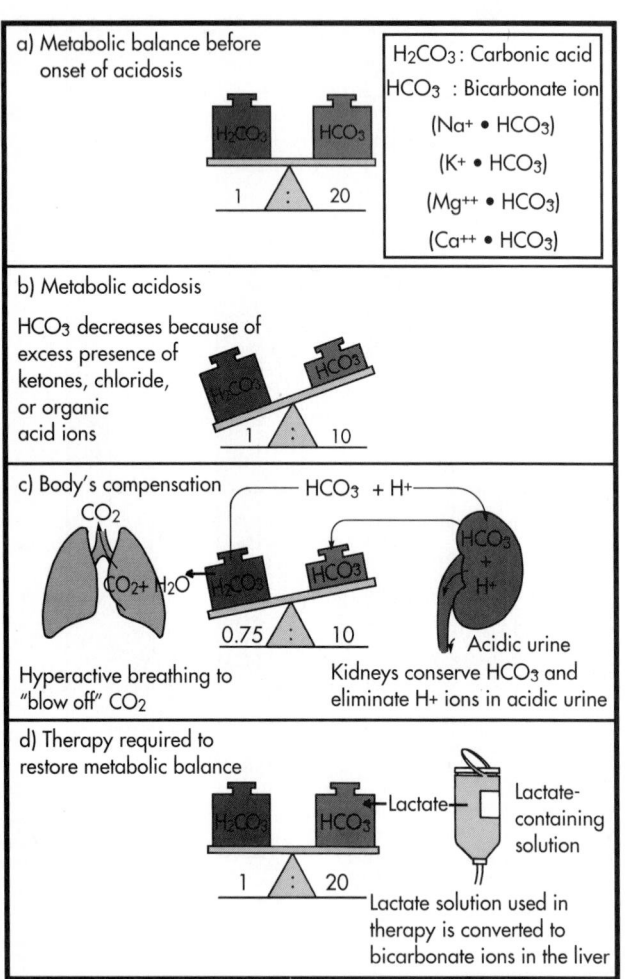

a) Metabolic balance before onset of acidosis

H_2CO_3 : Carbonic acid
HCO_3 : Bicarbonate ion
$(Na^+ \bullet HCO_3)$
$(K^+ \bullet HCO_3)$
$(Mg^{++} \bullet HCO_3)$
$(Ca^{++} \bullet HCO_3)$

1 : 20

b) Metabolic acidosis

HCO_3 decreases because of excess presence of ketones, chloride, or organic acid ions

1 : 10

c) Body's compensation

$HCO_3 + H^+$

CO_2

$CO_2 + H_2O$

0.75 : 10

Acidic urine

Hyperactive breathing to "blow off" CO_2

Kidneys conserve HCO_3 and eliminate H^+ ions in acidic urine

d) Therapy required to restore metabolic balance

Lactate

Lactate-containing solution

1 : 20

Lactate solution used in therapy is converted to bicarbonate ions in the liver

Metabolic acidosis.

136

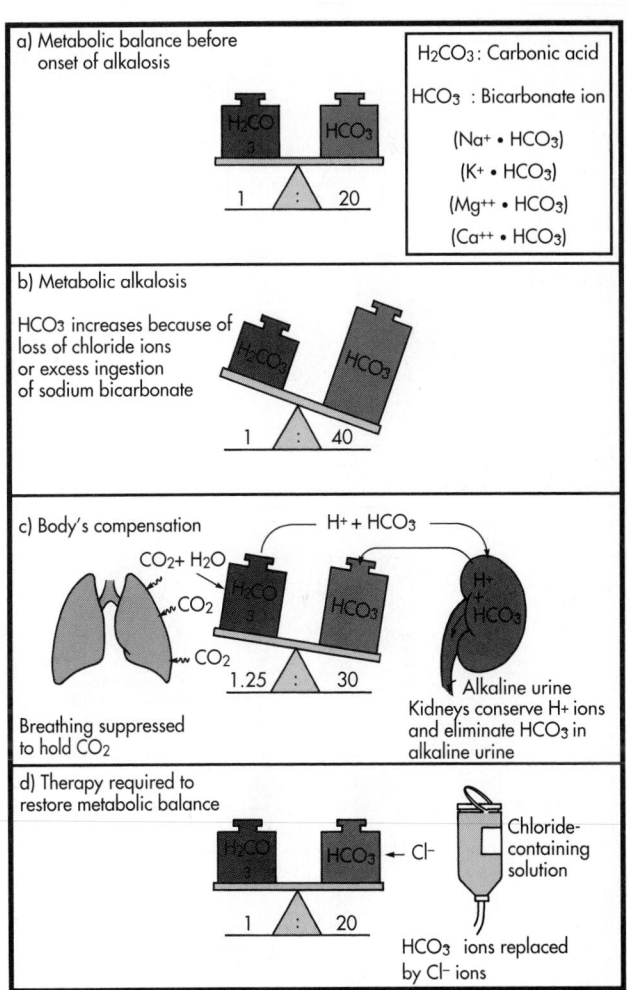

a) Metabolic balance before onset of alkalosis

H_2CO_3 : Carbonic acid

HCO_3 : Bicarbonate ion

$(Na^+ \cdot HCO_3)$

$(K^+ \cdot HCO_3)$

$(Mg^{++} \cdot HCO_3)$

$(Ca^{++} \cdot HCO_3)$

1 : 20

b) Metabolic alkalosis

HCO_3 increases because of loss of chloride ions or excess ingestion of sodium bicarbonate

1 : 40

c) Body's compensation

$H^+ + HCO_3$

$CO_2 + H_2O$

CO_2

CO_2

1.25 : 30

Breathing suppressed to hold CO_2

Alkaline urine Kidneys conserve H^+ ions and eliminate HCO_3 in alkaline urine

d) Therapy required to restore metabolic balance

1 : 20

Chloride-containing solution

$\leftarrow Cl^-$

HCO_3 ions replaced by Cl^- ions

Metabolic alkalosis.

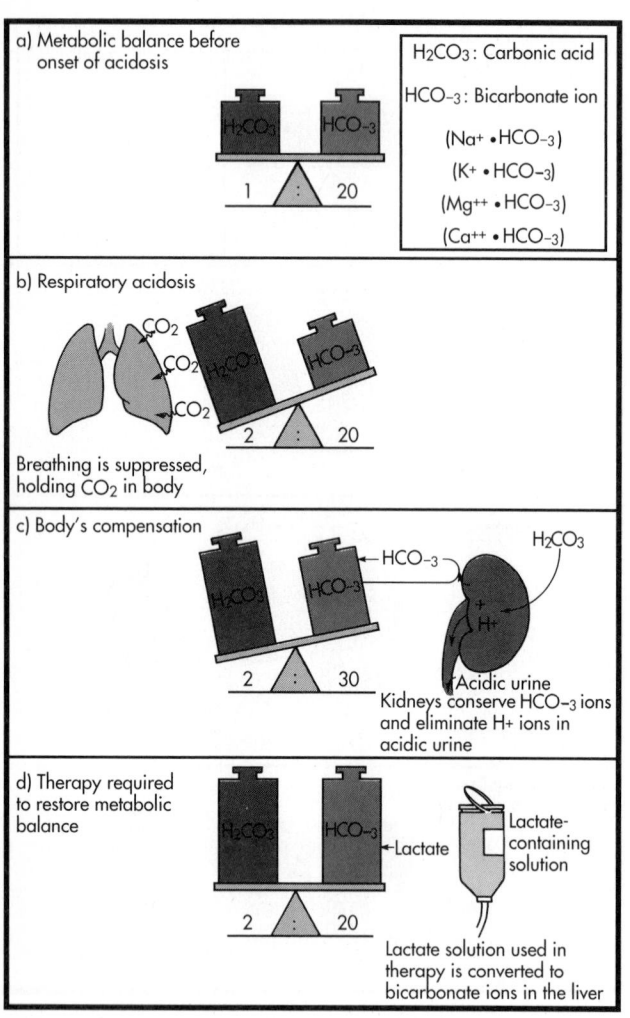

Respiratory acidosis.

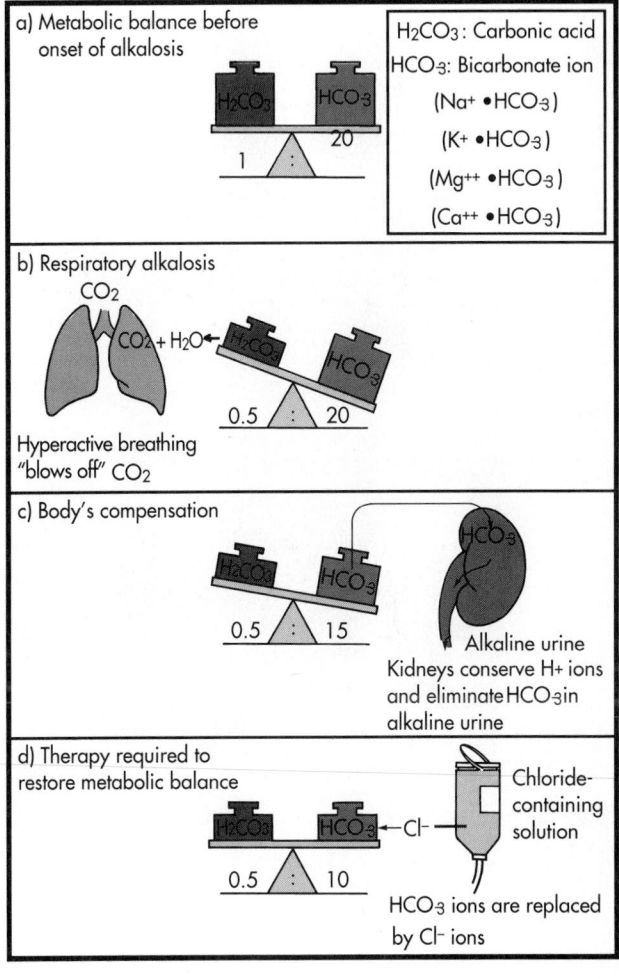

a) Metabolic balance before onset of alkalosis

H_2CO_3: Carbonic acid
HCO_3: Bicarbonate ion
$(Na^+ \cdot HCO_3)$
$(K^+ \cdot HCO_3)$
$(Mg^{++} \cdot HCO_3)$
$(Ca^{++} \cdot HCO_3)$

1 : 20

b) Respiratory alkalosis

CO_2
$CO_2 + H_2O$
H_2CO_3
HCO_3

0.5 : 20

Hyperactive breathing "blows off" CO_2

c) Body's compensation

H_2CO_3
HCO_3
HCO_3

0.5 : 15

Alkaline urine
Kidneys conserve H+ ions and eliminate HCO_3 in alkaline urine

d) Therapy required to restore metabolic balance

H_2CO_3
HCO_3
Cl^-

0.5 : 10

Chloride-containing solution

HCO_3 ions are replaced by Cl^- ions

Respiratory alkalosis.

139

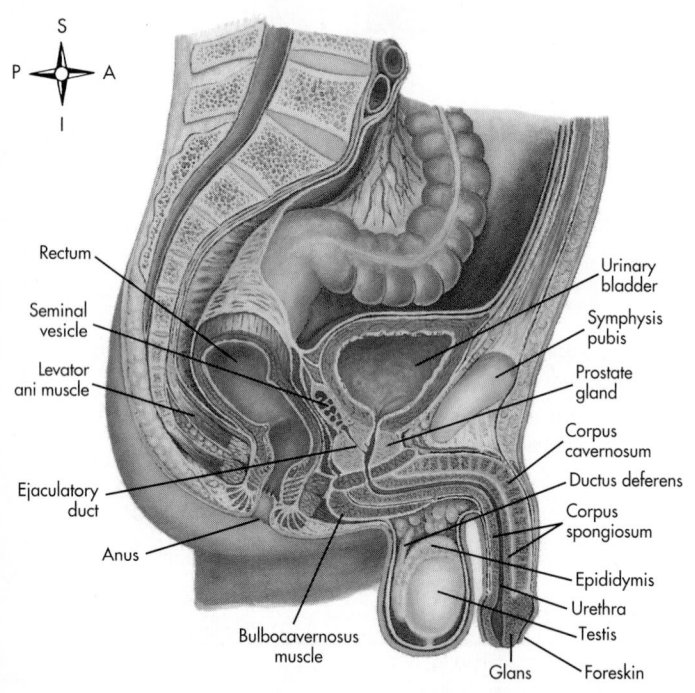

Male pelvic organs (lateral view of midsagittal section).

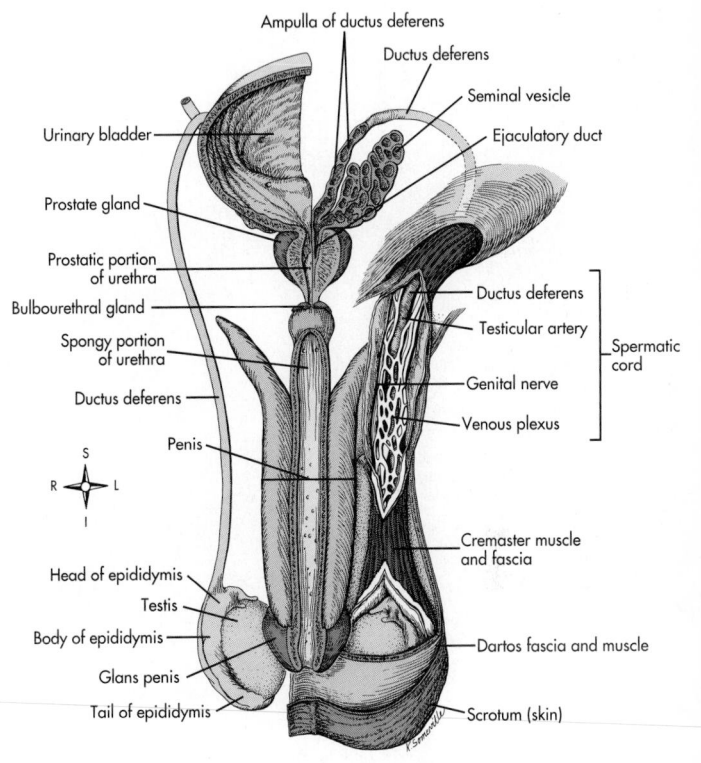

Ampulla of ductus deferens
Ductus deferens
Seminal vesicle
Ejaculatory duct
Urinary bladder
Prostate gland
Prostatic portion of urethra
Bulbourethral gland
Spongy portion of urethra
Ductus deferens
Penis
Ductus deferens
Testicular artery
Genital nerve
Venous plexus
Spermatic cord
Cremaster muscle and fascia
Head of epididymis
Testis
Body of epididymis
Glans penis
Tail of epididymis
Dartos fascia and muscle
Scrotum (skin)

Male reproductive system (anterior view).

Female pelvic organs
(lateral view of
midsagittal section).

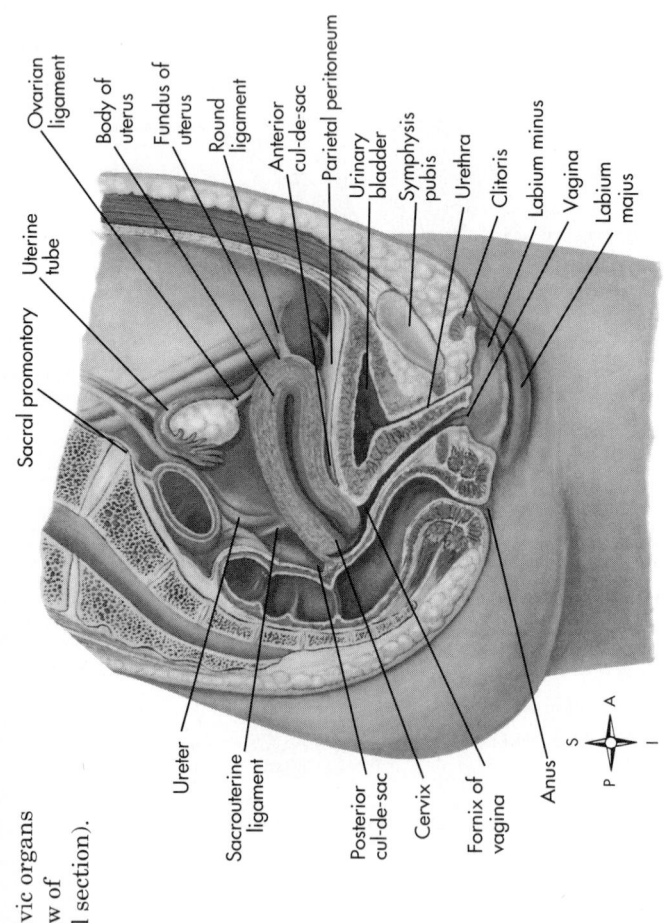

Ovarian
ligament

Body of
uterus

Fundus of
uterus

Round
ligament

Anterior
cul-de-sac

Parietal peritoneum

Urinary
bladder

Symphysis
pubis

Urethra

Clitoris

Labium minus

Vagina

Labium
majus

Uterine
tube

Sacral promontory

Ureter

Sacrouterine
ligament

Posterior
cul-de-sac

Cervix

Fornix of
vagina

Anus

142

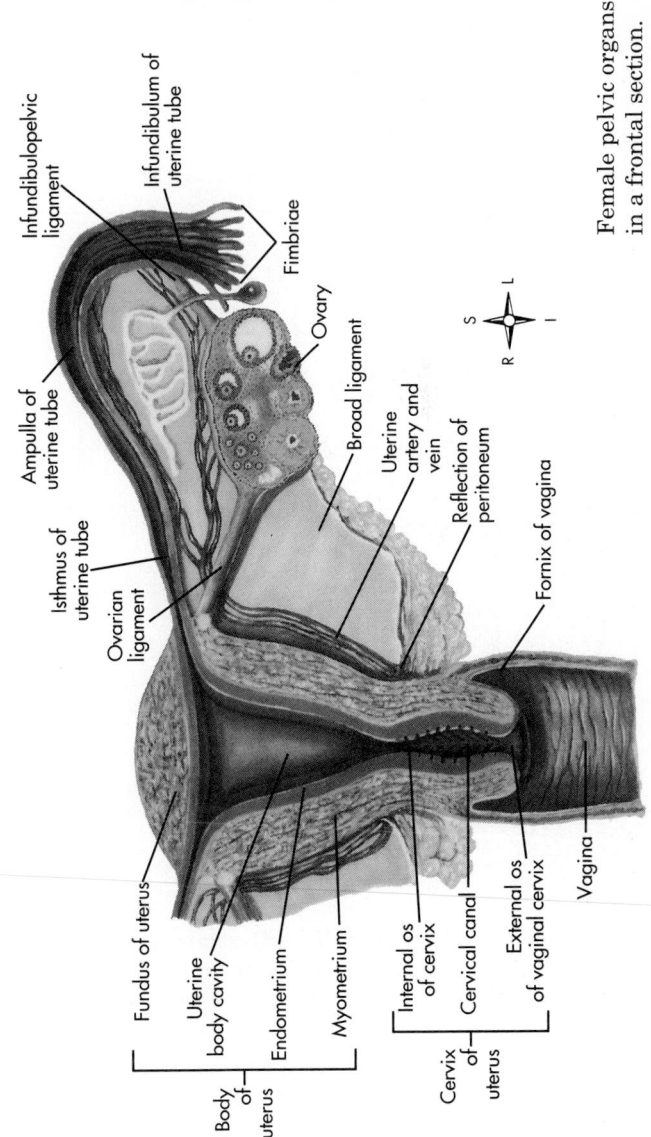

Female pelvic organs in a frontal section.

Infundibulopelvic ligament

Infundibulum of uterine tube

Fimbriae

Ampulla of uterine tube

Ovary

Isthmus of uterine tube

Broad ligament

Ovarian ligament

Uterine artery and vein

Reflection of peritoneum

Fornix of vagina

Fundus of uterus

Uterine body cavity

Endometrium

Myometrium

Body of uterus

Internal os of cervix

Cervical canal

External os of vaginal cervix

Cervix of uterus

Vagina

S
R — L
I

143

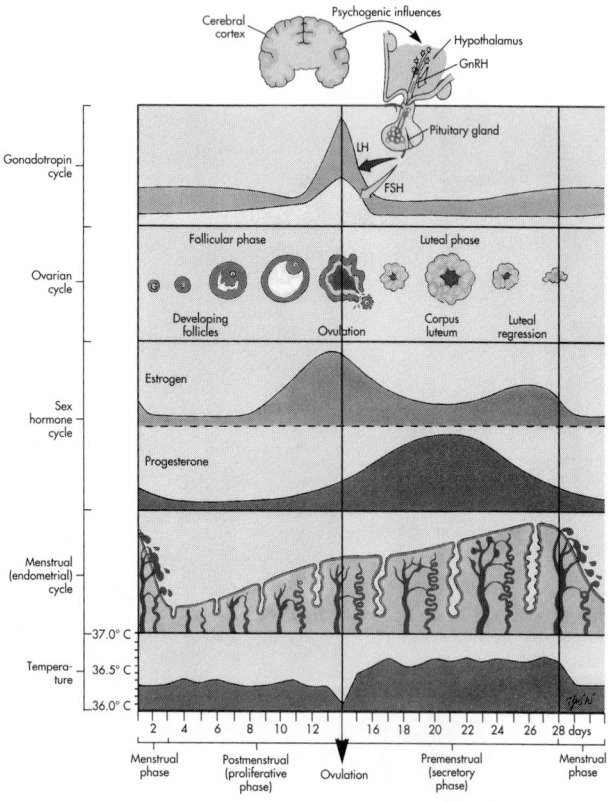

Human menstrual cycle. Diagram illustrates the relationship of pituitary, ovarian, and uterine functions throughout a usual 28-day cycle. A sharp increase in LH levels causes ovulation, whereas menstruation (sloughing off of the endometrial lining) is initiated by lower levels of progesterone.

BLOOD, PLASMA, AND SERUM VALUES

Test	Normal Values*	Significance of a Change
Acid phosphatase	*Women:* 0.01-0.56 sigma U/ml *Men:* 0.13-0.63 sigma U/ml	↑ in prostate cancer ↑ in kidney disease ↑ after trauma and in fever
Alkaline phosphatase	*Adult:* 13-39 IU/l *Child:* up to 104 IU/l	↑ in bone disorders ↑ in liver disease ↑ during pregnancy ↑ in hypothyroidism
Bicarbonate	22-26 mEq/L	↑ in metabolic alkalosis ↑ in respiratory alkalosis ↑ in metabolic acidosis ↑ in respiratory alkalosis
Blood urea nitrogen (BUN)	5-25 mg/100 ml	↑ with increased protein intake ↓ in kidney failure
Blood volume	*Women:* 65 ml/kg body weight *Men:* 69 ml/kg body weight	↓ during hemorrhage
Calcium	8.4-10.5 mg/100 ml	↑ in hypervitaminosis D ↑ in hyperparathyroidism ↑ in bone cancer and other bone diseases ↓ in severe diarrhea

*Values vary with the analysis method used.

(continued)

145

BLOOD, PLASMA, AND SERUM VALUES—cont'd

Test	Normal Values*	Significance of a Change
Calcium—cont'd		→ in hypoparathyroidism
		→ in avitaminosis D (rickets and osteomalacia)
Carbon dioxide content	24-32 mEq/L	↑ in severe vomiting
		↑ in respiratory disorders
		↑ in obstruction of intestines
		→ in acidosis
		→ in severe diarrhea
		→ in kidney disease
Chloride	96-107 mEq/L	↑ in hyperventilation
		↑ in kidney disease
		↑ in Cushing's syndrome
		→ in diabetic acidosis
		→ in severe diarrhea
		→ in severe burns
		→ in Addison's disease
Clotting time	5-10 minutes	→ in hemophilia
		→ (occasionally) in other clotting disorders
Copper	100-200 μg/100 ml	↑ in some liver disorders

Creatine phosphokinase (CPK)	Women: 5-35 mU/ml Men: 5-55 mU/ml	↑ in Duchenne's muscular dystrophy ↑ during myocardial infarction ↑ in muscle trauma
Creatine	0.06-1.5 mg/100 ml	↑ in some kidney disorders
Glucose	70-110 mg/100 ml (fasting)	↑ in diabetes mellitus ↑ in kidney disease ↑ in liver disease ↑ during pregnancy ↑ in hyperthyroidism → in hypothyroidism → in Addison's disease → in hyperinsulinism
Hematocrit (packed cell volume)	Women: 38%-47% Men: 40%-54%	↑ in polycythemia ↑ in severe dehydration → in anemia → in leukemia → in hyperthyroidism → in cirrhosis of liver
Hemoglobin	Women: 12-16g/100 ml Men: 13-18g/100 ml Newborn: 14-20g/100 ml	↑ in polycythemia ↑ in chronic obstructive pulmonary disease ↑ in congestive heart failure → in anemia → in hyperthyroidism → in cirrhosis of liver

(continued)

147

BLOOD, PLASMA, AND SERUM VALUES—cont'd

Test	Normal Values*	Significance of a Change
Iron	50-150 g/100 ml (can be higher in men)	↑ in liver disease ↑ in anemia (some forms) ↓ in iron-deficiency anemia
Lactic dehydrogenase (LDH)	60-120 U/ml	↑ during myocardial infarction ↑ in anemia (several forms) ↑ liver disease ↑ in acute leukemia and other cancers
Lipids—total Cholesterol—total High-density lipoprotein (HDL) Low-density lipoprotein (LDL)	450-1,000 mg/100 ml 120-220 mg/100 ml >40 mg/100 ml <180 mg/100 ml	↑ (total) in diabetes mellitus ↑ (total) in kidney disease ↑ (total) in hypothyroidism ↓ (total) in hyperthyroidism ↑ in inherited hypercholesterolemia ↑ (cholesterol) in chronic hepatitis ↓ (cholesterol) in acute hepatitis ↑ (HDL) with regular exercise
Triglycerides Phospholipids Fatty acids	40-150 mg/100 ml 145-200 mg/100 ml 190-420 mg/100 ml	
Mean corpuscular volume Osmolality	82-98 μl 285-295 mOsm/L	↑ or ↓ in various forms of anemia ↑ or ↓ in fluid and electrolyte imbalances

Test	Reference value	
P_{CO_2}	35-43 mm Hg	↑ in severe vomiting
		↑ in respiratory disorders
		↑ in obstruction of intestines
		↓ in acidosis
		↓ in severe diarrhea
		↓ in kidney disease
pH	7.35-7.45	↑ during hyperventilation
		↑ in Cushing's syndrome
		↓ during hypoventilation
		↓ in acidosis
		↓ in Addison's disease
Phosphorus	2.5-4.5 mg/100 ml	↑ in hypervitaminosis D
		↑ in kidney disease
		↑ in hypoparathyroidism
		↑ in acromegaly
		↓ hyperparathyroidism
		↓ in hypovitaminosis D (rickets and osteomalacia)
Plasma volume	Women: 40 ml/kg body weight	↑ or ↓ in fluid and electrolyte imbalances
	Men: 39 ml/kg body weight	↓ during hemorrhage
Platelet count	150,000-400,000/mm³	↑ in heart disease
		↑ in cancer
		↑ in cirrhosis of liver
		↑ after trauma
		↓ in anemia (some forms)

(continued)

BLOOD, PLASMA, AND SERUM VALUES—cont'd

Test	Normal Values*	Significance of a Change
Platelet count—cont'd		$\rightarrow$ during chemotherapy $\rightarrow$ in some allergies
P_{O_2}	75-100 mm Hg (breathing standard air)	$\uparrow$ in polycythemia $\rightarrow$ in anemia $\rightarrow$ in chronic obstructive pulmonary disease
Potassium	3.8 to 5.1 mEq/L	$\uparrow$ in hypoaldosteronism $\uparrow$ in acute kidney failure $\rightarrow$ in vomiting or diarrhea $\rightarrow$ in starvation
Protein—total Albumin Globulin	6-8.4 g/100 ml 3.5-5 g/100 ml 2.3-3.5 g/100 ml	$\uparrow$ (total) in severe dehydration $\rightarrow$ (total) during hemorrhage $\rightarrow$ (total) in starvation
Red blood cell count	*Women:* 4.2-5.4 million/mm³ *Men:* 4.5-6.2 million /mm³	$\uparrow$ in polycythemia $\uparrow$ in dehydration $\rightarrow$ in anemia (several forms) $\rightarrow$ in Addison's disease $\rightarrow$ in systemic lupus erythematosus
Reticulocyte count	25,000-75,000/mm³ (0.5%-1.5% of RBC count)	$\uparrow$ in hemolytic anemia $\uparrow$ in leukemia and metastatic carcinoma $\rightarrow$ in pernicious anemia

Sodium	136-145 mEq/L	↓ in iron-deficiency anemia
		↓ during radiation therapy
		↑ in dehydration
		↑ in trauma or disease of the central nervous system
		↑ or ↓ in kidney disorders
		↓ in excessive sweating, vomiting, diarrhea
		↓ in burns (sodium shift into cells)
Specific gravity	1.058	↑ or ↓ in fluid imbalances
Transaminase	10-40 U/ml	↑ during myocardial infarction
		↑ in liver disease
Uric acid	Women: 2.5-7.5 mg/100 ml	↑ in gout
	Men: 3-9 mg/100 ml	↑ in toxemia of pregnancy
		↑ during trauma
Viscosity	1.4-1.8 times the viscosity of water	↑ in polycythemia
		↑ in dehydration
White blood cell count		
Total	4,500-11,000 mm³	↑ (total) in acute infections
Neutrophils	50%-70% of total	↑ (total) in trauma
Eosinophils	2%-4% of total	↑ (total) some cancers
Basophils	0.5%-1% of total	↓ (total) in anemia (some forms)
Lymphocytes	20%-25% of total	↓ (total) during chemotherapy
Monocytes	7%-8% of total	↑ (neutrophil) in acute infection
		↓ (neutrophil) in allergies
		↑ (basophil) in severe allergies
		↑ (lymphocyte) during antibody reactions
		↑ (monocyte) in chronic infections

URINE COMPONENTS

Test	Normal Values*	Significance of a Change
Routine urinalysis		
Acetone and acetoacetate	0	↑ during fasting ⇑ in diabetic acidosis
Albumin	0-trace	↑ in hypertension ↑ in kidney disease ↑ after strenuous exercise (temporary)
Ammonia	20-70 mEq/L	↑ in liver disease ↑ in diabetes mellitus
Bile and bilirubin	—	↑ during obstruction of the bile ducts
Calcium	<150 mg/day	↑ in hyperparathyroidism ↓ in hypoparathyroidism
Color	Transparent yellow, straw-colored, or amber	Abnormal color or cloudiness may indicate blood in urine, bile, bacteria, drugs, food pigments, or high solute concentration
Odor	Characteristic slight odor	Acetone odor in diabetes mellitus (diabetic ketosis)
Osmolality	500-800 mOsm/L	↑ in dehydration ↑ in heart failure ↓ in diabetes insipidus ↓ in aldosteronism

pH	4.6-8.0	↑ in alkalosis ↑ during urinary infections ↓ in acidosis ↓ in dehydration ↓ in emphysema
Potassium	25-100 mEq/L	↑ in dehydration ↑ in chronic kidney failure ↓ in diarrhea or vomiting ↓ in adrenal insufficiency
Sodium	75-200 mg/day	↑ in starvation ↑ in dehydration ↓ in acute kidney failure ↓ in Cushing's syndrome
Creatinine clearance	100-140 ml/min	↑ in kidney disease
Creatine	1-2 g/day	↑ in infections ↓ in some kidney diseases ↓ in anemia (some forms)
Glucose	0	↑ in diabetes mellitus ↑ in hyperthyroidism ↑ in hypersecretion of adrenal cortex
Urea clearance	>40 ml blood cleared per min	↑ in some kidney diseases
Urea	25-35 g/day	↑ in some liver diseases ↑ in hemolytic anemia

*Values vary with the analysis method used.

(continued)

URINE COMPONENTS—cont'd

Test	Normal Values*	Significance of a Change
Urea—cont'd		↓ during obstruction of bile ducts ↓ in severe diarrhea
Uric acid	0.6-1.0 g/day	↑ in gout ↓ in some kidney diseases
Microscopic examination		
Bacteria	<10,000/ml	↑ during urinary infections
Blood cells (RBC)	0-trace	↑ in pyelonephritis ↑ from damage by calculi ↑ in infection ↑ in cancer
Blood cells (WBC)	0-trace	↑ in infections
Blood cell casts (RBC)	0-trace	↑ in pyelonephritis
Blood cell casts (WBC)	0-trace	↑ in infection
Crystals	0-trace	↑ in urinary retention Very large crystalline masses are calculi
Epithelial casts	0-trace	↑ in some kidney disorders ↑ in heavy metal toxicity
Granular casts	0-trace	↑ in in some kidney disorders
Hyaline casts	0-trace	↑ in some kidney disorders ↑ in fever

CONVERSION FACTORS (TO SI UNITS)

Component	Normal Range in Units as Customarily Reported	Conversion Factor	Normal Range in SI Units, Molecular Units International Units, or Decimal Fractions
Biochemical Components of Blood*			
Acetoacetic acid (S)	0.2-1.0 mg/dL	98	19.6-98.0 μmol/L
Acetone (S)	0.3-2.0 mg/dL	172	51.6-344.0 μmol/L
Albumin (S)	3.2-4.5 g/dL	10	32-45 g/L
Ammonia (P)	20-120 μg/dL	0.588	11.7-70.5 μmol/L
Amylase (S)	60-160 Somogyi units/dL	1.85	111-296 U/L
Base, total (S)	145-160 mEq/L	1	145-160 mmol/L
Bicarbonate (P)	21-28 mEq/L	1	21-28 mmol/L
Bile acids (S)	0.3-3.0 mg/dL	10	3-30 mg/L
		2.547	0.8-7.6 μmol/L
Bilirubin, direct (S)	Up to 0.3 mg/dL	17.1	Up to 5.1 μmol/L
Bilirubin, indirect (S)	0.1-1.0 mg/dL	17.1	1.7-17.1 μmol/L
Blood gases (B)			
PCO_2 arterial	35-40 mm Hg	0.133	4.66-5.32 kPa
PO_2 arterial	95-100 mm Hg	0.133	12.64-13.30 kPa
Calcium (S)	8.5-10.5 mg/dL	0.25	2.1-2.6 mmol/L
Chloride (S)	95-103 mEq/L	1	95-103 mmol/L

(continued)

CONVERSION FACTORS (TO SI UNITS)—cont'd

Component	Normal Range in Units as Customarily Reported	Conversion Factor	Normal Range in SI Units, Molecular Units International Units, or Decimal Fractions
Biochemical Components of Blood*—cont'd			
Creatine (S)	0.1-0.4 mg/dl	76.3	7.6-30.5 μmol/L
Creatinine (S)	0.6-1.2 mg/dl	88.4	53-106 μmol/L
Creatinine clearance (P)	107-139 mL/min	0.0167	1.78-2.32 mL/s
Fatty acids (total) (S)	8-20 mg/dl	0.01	0.08-2.00 mg/L
Fibrinogen (P)	200-400 mg/dl	0.01	2.00-4.00 g/L
Gamma globulin (S)	0.5-1.6 g/dl	10	5-16 g/L
Globulins (total) (S)	2.3-3.5 g/dl	10	23-35 g/L
Glucose (fasting) (S)	70-110 mg/dl	0.055	3.85-6.05 mmol/L
Insulin (radioimmunoassay) (P)	4-24 μIU/ml	0.0417	0.17-1.00 μg/L
	0.20-0.84 μg/L	172.2	35-145 pmol/L
Iodine, BEI (S)	3.5-6.5 μg/dl	0.079	0.280-0.51 μmol/L
Iodine, PBI (S)	4.0-8.0 μg/dl	0.079	0.320-0.63 μmol/L
Iron, total (S)	60-150 μg/dl	0.179	11-27 μmol/L
Iron-binding capacity (S)	300-360 μg/dl	0.179	54-64 μmol/L
17-Ketosteroids (P)	25-125 μg/dl	0.01	0.25-1.25 mg/L
Lactic dehydrogenase (S)	80-120 units at 30 °C	0.48	38-62 U/L at 30 °C

	Conventional	Factor	SI
Lactate → pyruvate	100-190 U/L at 37 °C	1	100-190 U/L at 37 °C
Lipase (S)	0-1.5 U/ml (Cherry-Crandall)	278	0-417 U/L
Lipids (total) (S)	400-800 mg/dl	0.01	4.00-8.00 g/L
Cholesterol	150-250 mg/dl	0.026	3.9-6.5 mmol/L
Triglycerides	75-165 mg/dl	0.0114	0.85-1.89 mmol/L
Phospholipids	150-380 mg/dl	0.01	1.50-380 g/L
Free fatty acids	9.0-15.0 mM/L	1	9.0-15.0 mmol/L
Nonprotein nitrogen (S)	20-35 mg/dl	0.714	14.3-25.0 mmol/L
Phosphatase (P)			
Acid (units/dL)			
Cherry-Crandall		2.77	0-5.5 U/L
King-Armstrong		1.77	0-5.5 U/L
Bodansky		5.37	0-5.5 U/L
Alkaline (units/dL)			
King-Armstrong		1.77	30-120 U/L
Bodansky		5.37	30-120 U/L
Bessey-Lowry-Brock		16.67	30-120 U/L
Phosphorus, inorganic (S)	3.0-4.5 mg/dl	0.323	0.97-1.45 mmol/L
Potassium (P)	3.8-5.0 mEq/L	1	3.8-5.0 mmol/L
Proteins, total (S)	6.0-7.8 g/dl	10	60-78 g/L
Albumin	3.2-4.5 g/dl	10	32-45 g/L
Globulin	2.3-3.5 g/dl	10	23-35 g/L
Sodium (P)	136-142 mEq/L	1	136-142 mmol/L
Testosterone: Male (S)	300-1,200 ng/dl	0.035	10.5-42.0 nmol/L
Female	30-95 ng/dl	0.035	1.0-3.3 nmol/L

(continued)

CONVERSION FACTORS (TO SI UNITS)—cont'd

Component	Normal Range in Units as Customarily Reported	Conversion Factor	Normal Range in SI Units, Molecular Units International Units, or Decimal Fractions
Biochemical Components of Blood*—cont'd			
Thyroid tests (S)			
Thyroxine (T_4)	4-11 μg/dL	12.87	51-142 nmol/L
T_4 expressed as iodine	3.2-7.2 μg/dL	79.0	253-569 nmol/L
T_3 resin uptake	25%-38% relative uptake	0.01	0.25%-9.38% relative uptake
TSH (S)	10 μU/mL	1	$<10^{-3}$ IU/L
Urea nitrogen (S)	8-23 mg/dL	0.357	2.9-8.2 mmol/L
Uric acid (S)	2-6 mg/dL	59.5	0.120-0.360 mmol/L
Vitamin B_{12} (S)	160-195 pg/mL	0.74	118-703 pmol/L
Hematology Values*			
Red cell volume (male)	25-35 mL/kg body weight	0.001	0.025-0.035 L/kg body weight
Hematocrit	40%-50%	0.01	0.40-0.50
Hemoglobin	13.5-18.0 g/dL	10	135-180 g/L
Hemoglobin	13.5-18.0 g/dL	0.155	2.09-2.79 mmol/L
RBC count	4.5-6 $\times 10^6/\mu$L	1	4.6-6 $\times 10^{12}$/L
WBC count	4.5-10 $\times 10^3/\mu$L	1	4.5-10 $\times 10^9$/L
Mean corpuscular volume	80-96 μm^3	1	80-96 fL

*The International Committee for Standardization in Hematology recommends that the numbers remain the same but that the units change, so that hemoglobin is expressed as grams per deciliter (g/dL) even though other measurements are expressed as units per liter (U/L).

158

Mini-Atlas

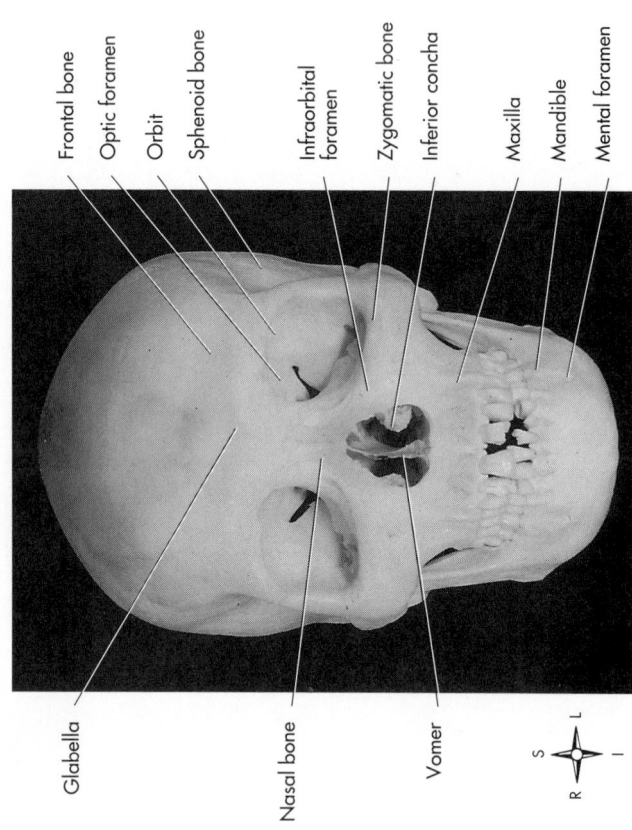

Frontal bone
Optic foramen
Orbit
Sphenoid bone
Infraorbital foramen
Zygomatic bone
Inferior concha
Maxilla
Mandible
Mental foramen

Glabella
Nasal bone
Vomer

S
R — L
I

Anterior view
of the skull.

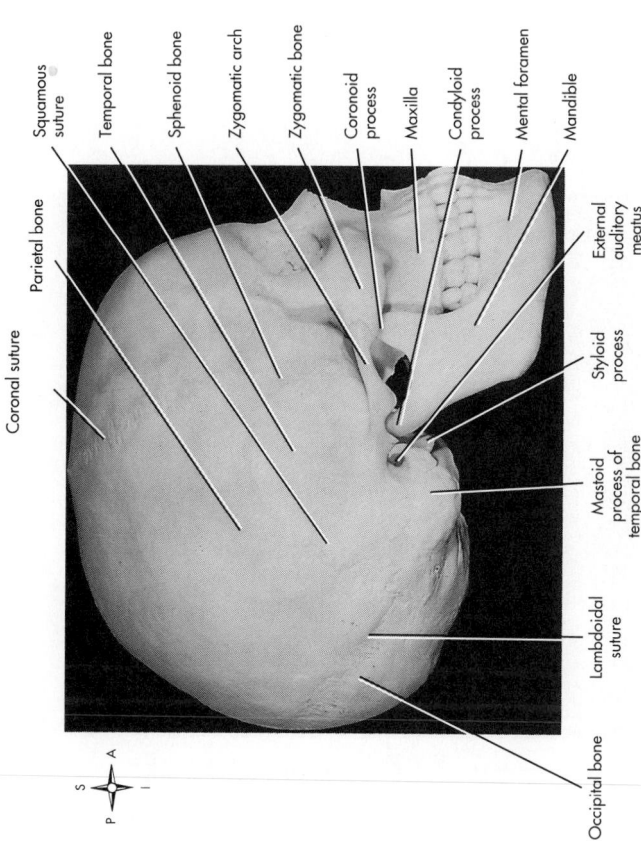

Skull viewed from the right side.

Coronal suture

Parietal bone

Squamous suture

Temporal bone

Sphenoid bone

Zygomatic arch

Zygomatic bone

Coronoid process

Maxilla

Condyloid process

Mental foramen

Mandible

Occipital bone

Lambdoidal suture

Mastoid process of temporal bone

Styloid process

External auditory meatus

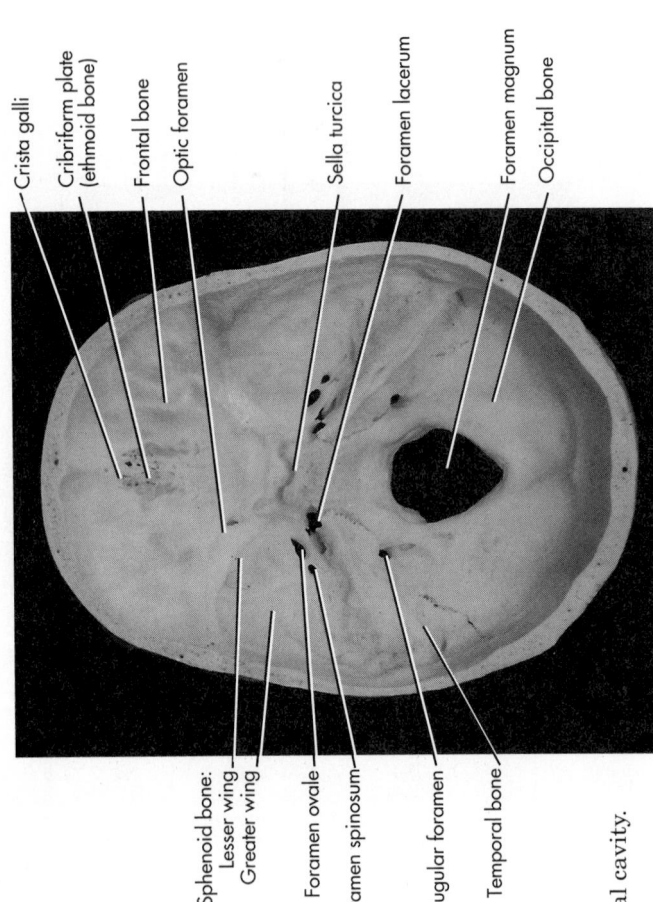

Crista galli

Cribriform plate (ethmoid bone)

Frontal bone

Optic foramen

Sella turcica

Foramen lacerum

Foramen magnum

Occipital bone

Sphenoid bone:
Lesser wing
Greater wing

Foramen ovale

Foramen spinosum

Jugular foramen

Temporal bone

Floor of the cranial cavity.

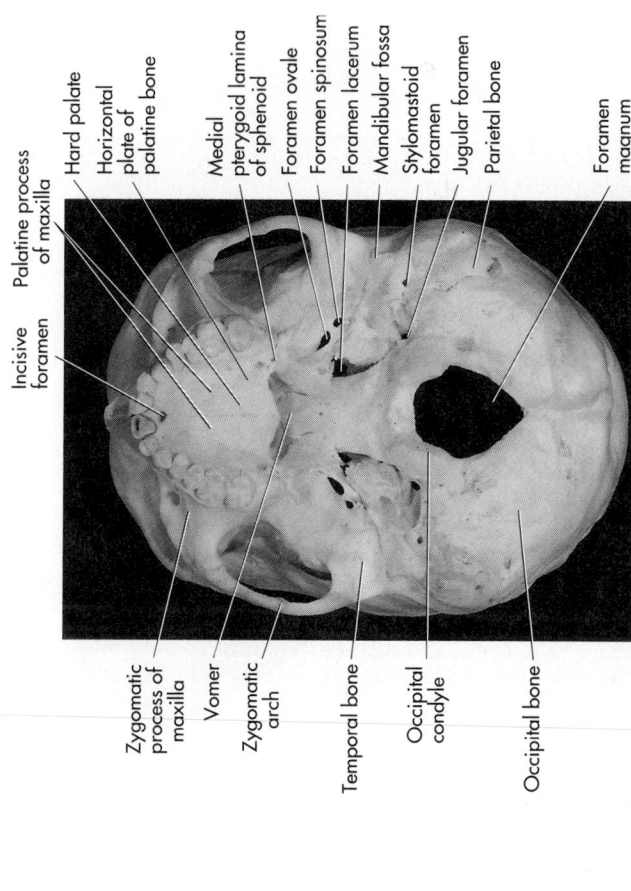

Incisive foramen

Palatine process of maxilla

Hard palate

Horizontal plate of palatine bone

Medial pterygoid lamina of sphenoid

Foramen ovale

Foramen spinosum

Foramen lacerum

Mandibular fossa

Stylomastoid foramen

Jugular foramen

Parietal bone

Foramen magnum

Zygomatic process of maxilla

Vomer

Zygomatic arch

Temporal bone

Occipital condyle

Occipital bone

Skull viewed from below.

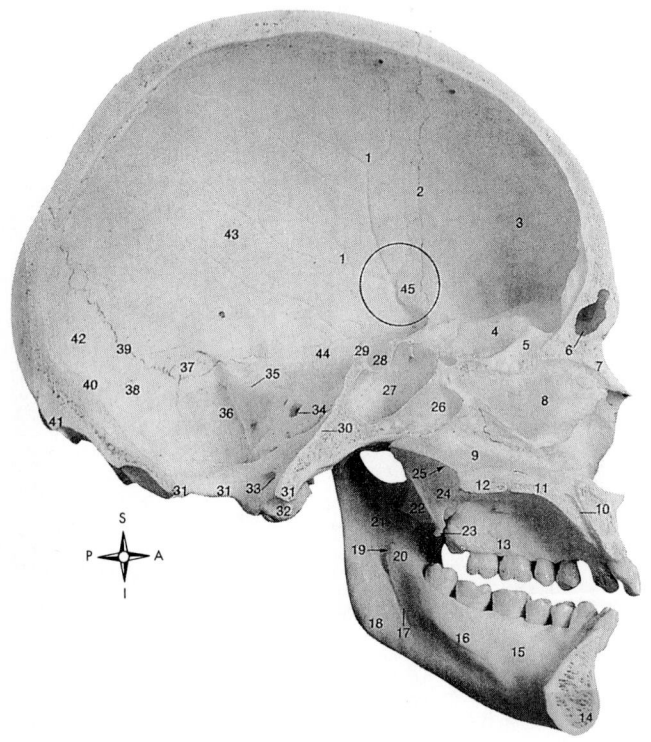

Left half of skull. Sagittal section.

1 Grooves for middle meningeal vessels
2 Coronal suture
3 Squamous part of frontal bone
4 Orbital part of frontal bone
5 Crista galli of ethmoid bone
6 Frontal sinus
7 Nasal bone
8 Perpendicular plate of ethmoid bone
9 Vomer
10 Incisive canal
11 Palatine process of maxilla
12 Horizontal plate of palatine bone
13 Alveolar process of maxilla
14 Mental protuberance
15 Body of mandible
16 Mylohyoid line
17 Groove for mylohyoid nerve
18 Angle of mandible
19 Mandibular foramen
20 Lingula
21 Ramus of mandible
22 Lateral pterygoid plate
23 Pterygoid hamulus of medial pterygoid plate
24 Medial pterygoid plate
25 Posterior nasal aperture (choana)
26 Right sphenoidal sinus
27 Left sphenoidal sinus
28 Pituitary fossa (sella turcica)
29 Dorsum sellae
30 Clivus
31 Margin of foramen magnum
32 Occipital condyle
33 Hypoglossal canal
34 Internal acoustic meatus in petrous part of temporal bone
35 Groove for superior petrosal sinus
36 Groove for sigmoid sinus
37 Mastoid (posterior inferior) angle of parietal bone
38 Groove for transverse sinus
39 Lambdoidal suture
40 Internal occipital protuberance
41 External occipital protuberance
42 Occipital bone
43 Parietal bone
44 Squamous part of temporal bone
45 Pterion (encircled)

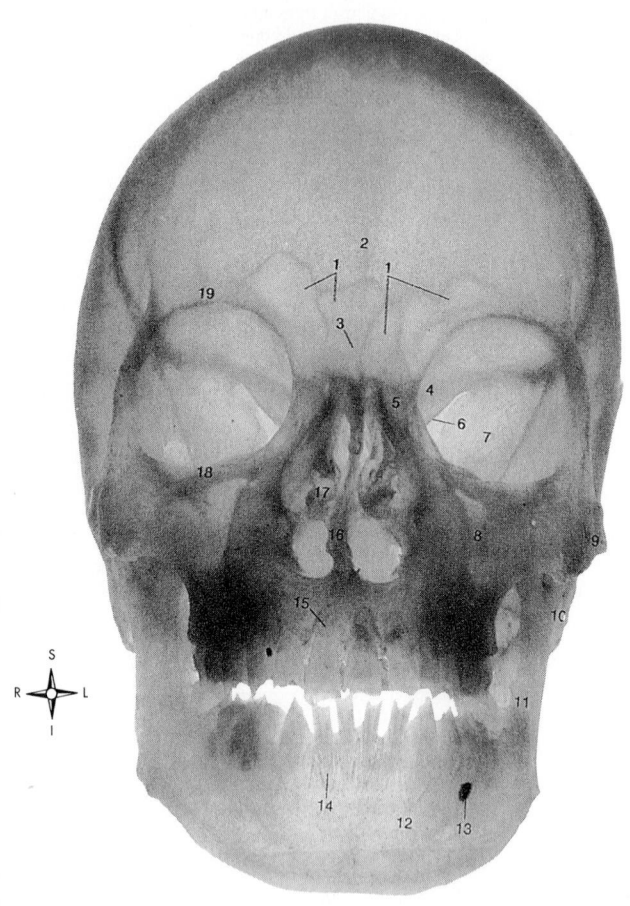

Skull, from the front. Cleared specimen, illuminated from
behind.

1	Frontal sinus	11	Ramus ⎫ of mandible
2	Frontal crest	12	Body ⎭
3	Crista galli	13	Mental foramen
4	Lesser wing of sphenoid	14	Root of lateral incisor
5	Ethmoidal sinus	15	Root of central incisor
6	Superior orbital fissure	16	Nasal septum
7	Greater wing of sphenoid	17	Inferior nasal concha
8	Maxillary sinus	18	Infraorbital margin
9	Zygomatic arch	19	Supraorbital margin
10	Mastoid process		

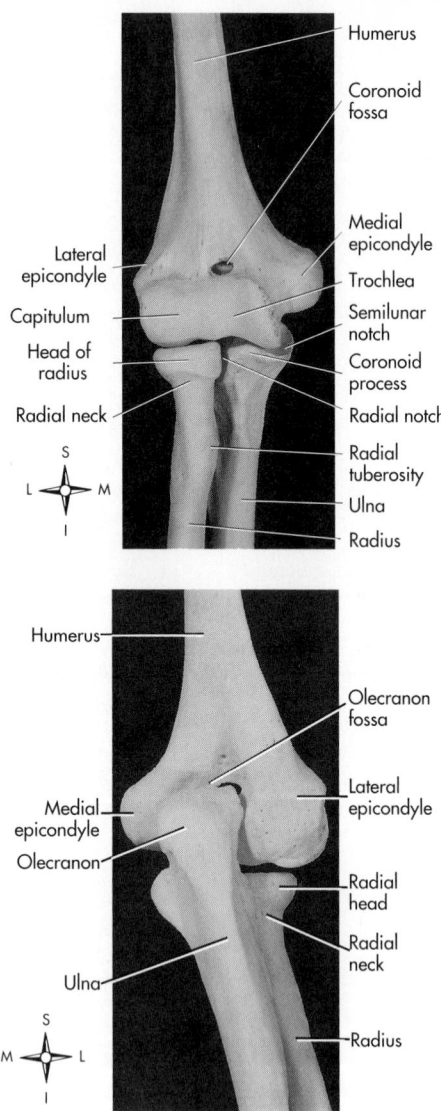

Bones of the elbow (*top,* anterior view; *bottom,* posterior view).

168

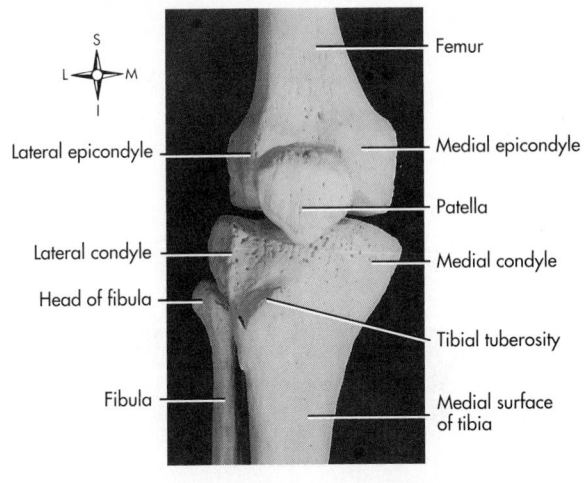

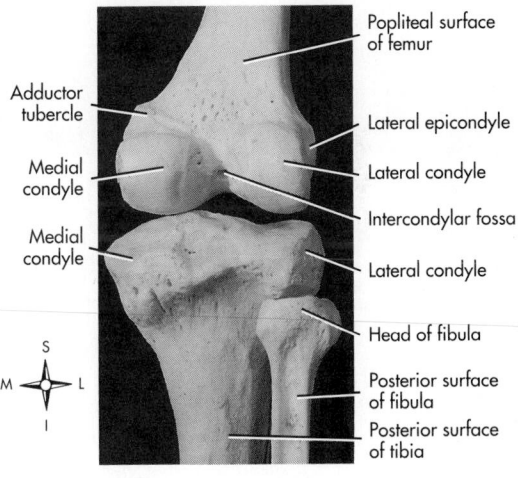

Bones of the knee (*top,* anterior view; *bottom,* posterior view).

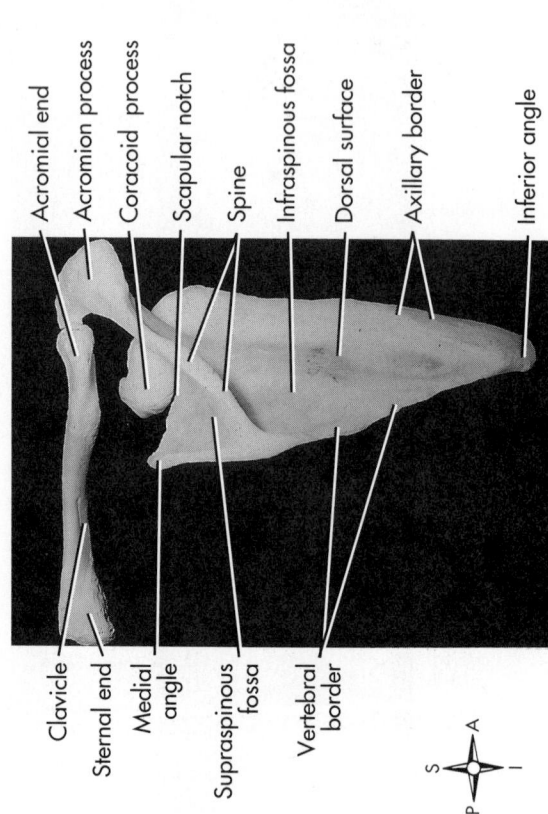

Acromial end
Acromion process
Coracoid process
Scapular notch
Spine
Infraspinous fossa
Dorsal surface
Axillary border
Inferior angle

Clavicle
Sternal end
Medial angle
Supraspinous fossa
Vertebral border

S
P — A
I

Bones of the shoulder; posterior view.

Bones of the pelvis; inferior view.

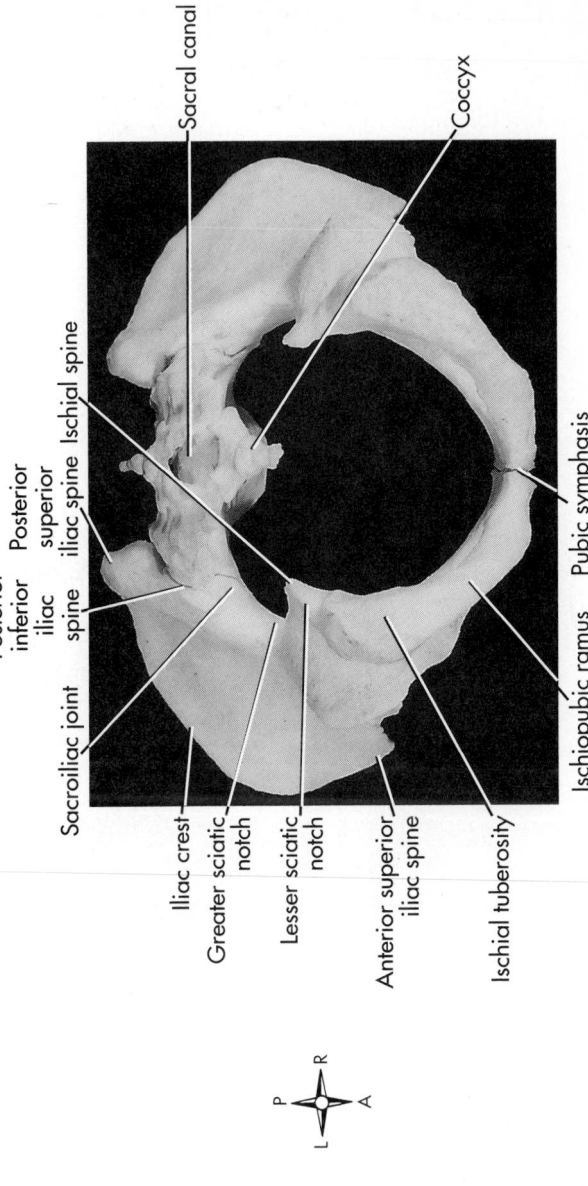

Sacral canal

Coccyx

Pubic symphasis

Ischiopubic ramus

Ischial tuberosity

Anterior superior iliac spine

Lesser sciatic notch

Greater sciatic notch

Iliac crest

Sacroiliac joint

Posterior inferior iliac spine

Posterior superior iliac spine

Ischial spine

171

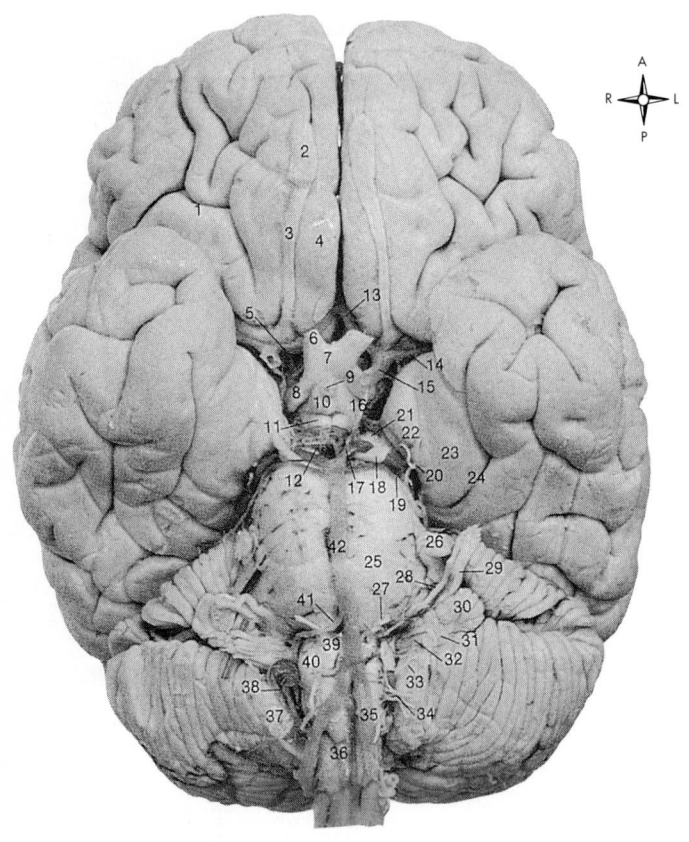

Ventral surface of the brain.

1	Orbital sulcus	23	Parahippocampal gyrus
2	Olfactory bulb	24	Collateral sulcus
3	Olfactory tract	25	Pons
4	Gyrus rectus	26	Trigeminal nerve
5	Anterior perforated substance	27	Abducens nerve
6	Optic nerve	28	Facial nerve
7	Optic chiasma	29	Vestibulocochlear nerve
8	Optic tract	30	Flocculus of cerebellum
9	Pituitary stalk (infundibulum)	31	Choroid plexus from lateral recess of fourth ventricle
10	Tuber cinereum and median eminence	32	Roots of glossopharyngeal, vagus and accessory nerves
11	Mamillary body	33	Spinal part of accessory nerve
12	Posterior perforated substance	34	Rootlets of hypoglossal nerve (superficial to marker)
13	Anterior cerebral artery	35	Vertebral artery
14	Middle cerebral artery	36	Medulla oblongata
15	Internal carotid artery	37	Tonsil of cerebellum
16	Posterior communicating artery	38	Posterior inferior cerebellar artery
17	Posterior cerebral artery	39	Pyramid ⎤ of medulla
18	Oculomotor nerve	40	Olive ⎦ oblongata
19	Superior cerebellar artery	41	Anterior inferior cerebellar artery
20	Trochlear nerve	42	Basilar artery
21	Crus of cerebral peduncle		
22	Uncus		

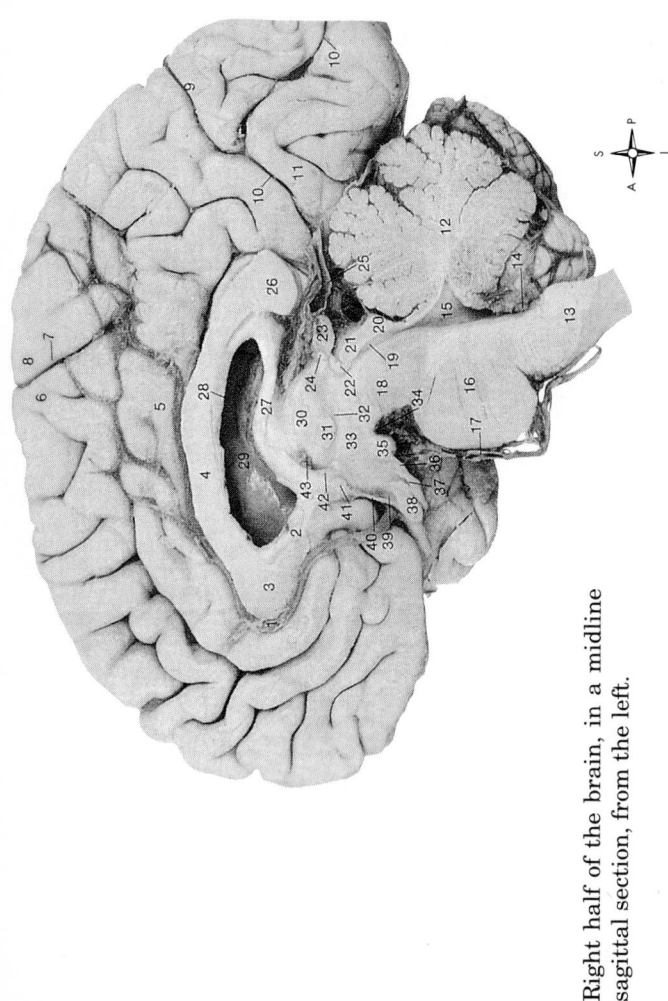

Right half of the brain, in a midline sagittal section, from the left.

175

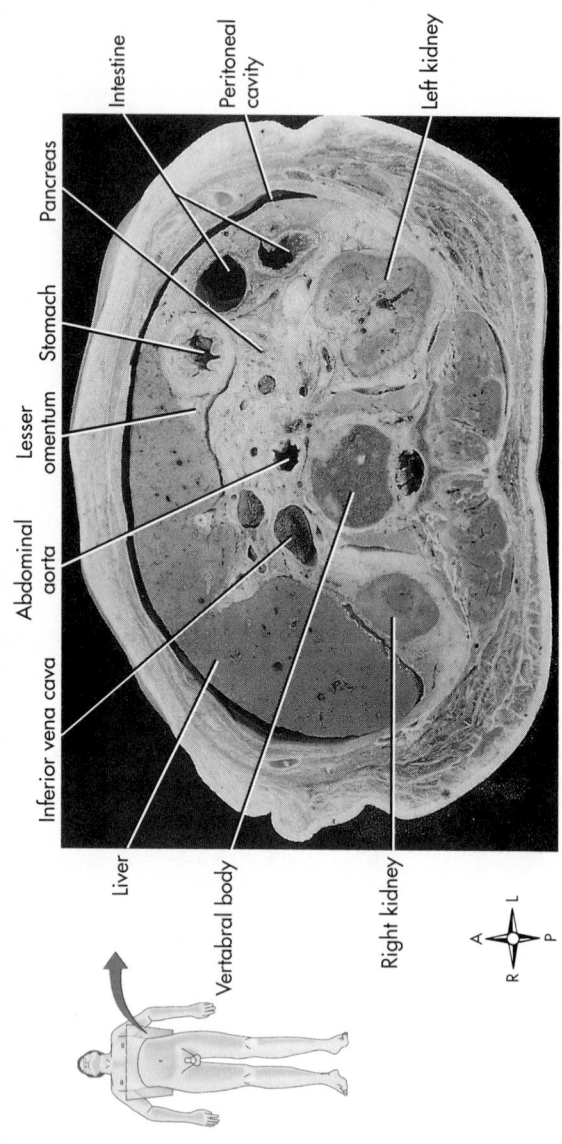

Intestine

Peritoneal cavity

Left kidney

Pancreas

Stomach

Lesser omentum

Abdominal aorta

Inferior vena cava

Liver

Vertebral body

Right kidney

A
R — L
P

Transverse section of the abdomen.

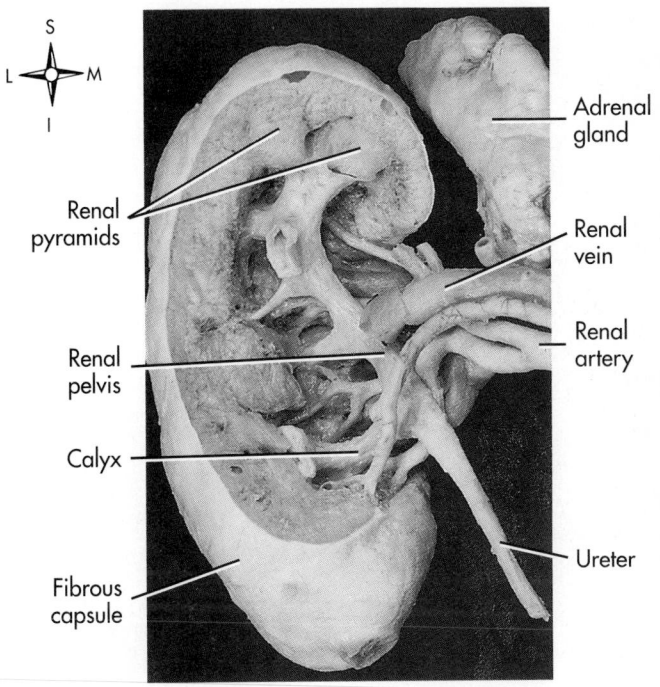

S
L — M
I

Adrenal
gland

Renal
pyramids

Renal
vein

Renal
artery

Renal
pelvis

Calyx

Ureter

Fibrous
capsule

Coronal section of the right kidney.

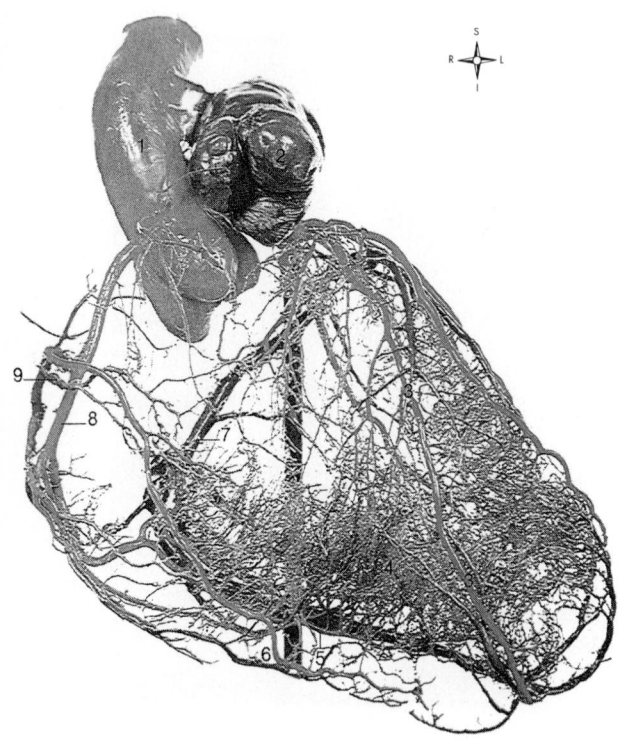

Cast of the cardiac vessels, from the front.

1 Ascending aorta
2 Pulmonary trunk and sinuses above pulmonary valve
  cusps
3 Anterior interventricular branch of left coronary artery
  and great cardiac vein
4 Vessels of interventricular septum
5 Middle cardiac vein and posterior interventricular
  branch of right coronary artery
6 Marginal branch of right coronary artery and small
  cardiac vein
7 Coronary sinus
8 Right coronary artery
9 Anterior cardiac vein

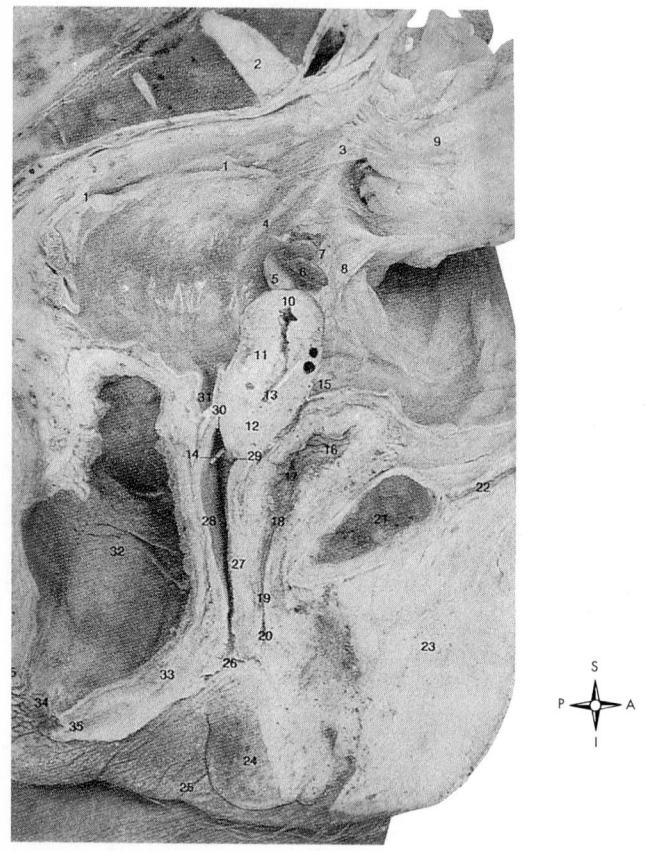

Left half of a midline sagittal section of the female pelvis.

1   Line of attachment of right limb of
    sigmoid mesocolon
2   Fifth lumbar intervertebral disk
3   Apex of sigmoid mesocolon
4   Ureter underlying peritoneum
5   Ovary
6   Uterine tube
7   Suspensory ligament of ovary
    containing ovarian vessels
8   Left limb of sigmoid mesocolon
    overlying external iliac vessels
9   Sigmoid colon (reflected to left
    and upwards)
10 Fundus ⎤
11 Body   ⎬ of uterus
12 Cervix ⎦
13 Marker in internal os
14 Marker in external os
15 Vesico-uterine pouch
16 Bladder
17 Marker in left ureteral orifice
18 Internal urethral orifice
19 Urethra
20 External urethral orifice
21 Pubic symphysis
22 Rectus abdominis (turned forwards)
23 Fat of mons pubis
24 Labium minus
25 Labium majus
26 Vestibule   ⎤
27 Anterior wall   ⎥
28 Posterior wall ⎬ of vagina
29 Anterior fornix ⎥
30 Posterior fornix⎦
31 Recto-uterine pouch
32 Rectum
33 Perineal body
34 Anal canal
35 External anal sphincter

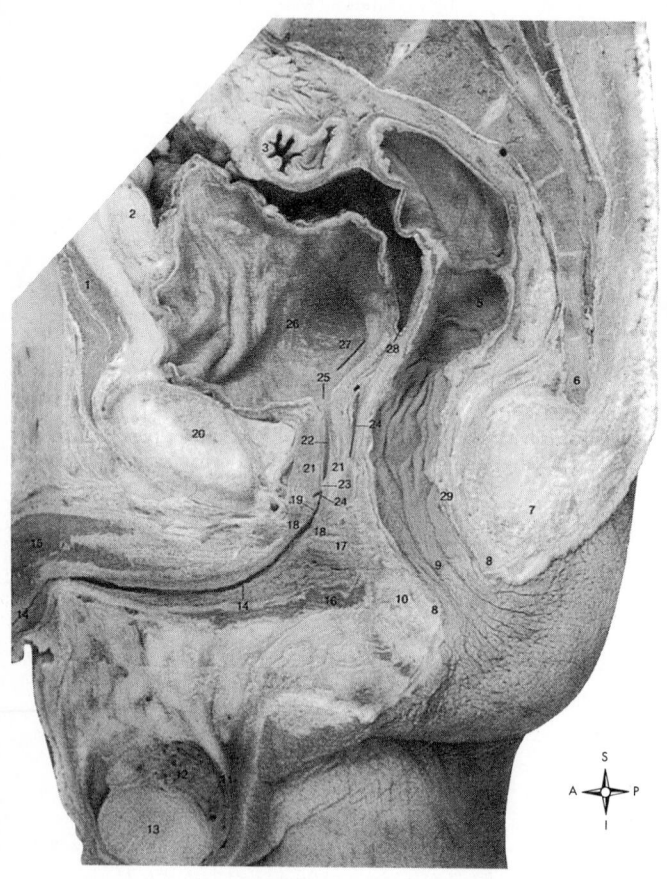

Right half of a midline sagittal section of the male pelvis.

1 Rectus abdominis
2 Extraperitoneal fat
3 Sigmoid colon
4 Promontory of sacrum
5 Rectum
6 Coccyx
7 Anococcygeal body
8 External anal sphincter
9 Anal canal with anal columns of mucous membrane
10 Perineal body
11 Ductus deferens
12 Epididymis
13 Testis
14 Spongy part of urethra and corpus spongiosum
15 Corpus cavernosum
16 Bulbospongiosus
17 Perineal membrane
18 Sphincter urethrae
19 Membranous part of urethra
20 Pubic symphysis
21 Prostate
22 Prostatic part of urethra
23 Seminal colliculus
24 Bristle in ejaculatory duct
25 Internal urethral orifice
26 Bladder
27 Bristle passing up into right ureteral orifice
28 Rectovesical pouch
29 Puborectalis fibers of levator ani

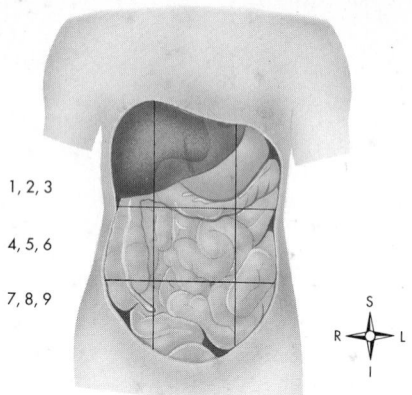

Nine regions of the abdominopelvic cavity. The nine regions of the abdominopelvic cavity showing the most superficial organs. 1, Right hypochondriac region. 2, Epigastric region. 3, Left hypochondriac region. 4, Right lumbar region. 5, Umbilical region. 6, Left lumbar region. 7, Right iliac region. 8, Hypogastric region. 9, Left iliac region.

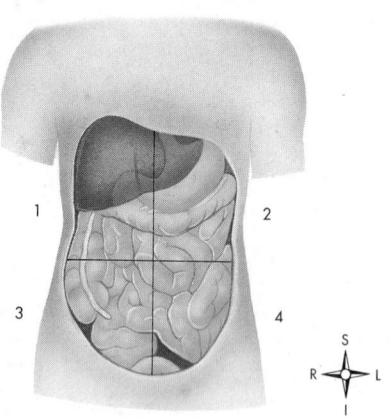

Division of the abdomen into four quadrants. Diagram shows relationship of internal organs to the four abdominopelvic quadrants: 1, right upper quadrant (RUQ); 2, left upper quadrant (LUQ); 3, right lower quadrant (RLQ); 4, left lower quadrant (LLQ).